THE TECHNICIAN WRITES

A Guide to Basic Technical Writing

ARNOLD B. SKLARE
Southampton College of Long Island University

BOYD & FRASER PUBLISHING COMPANY

Arnold B. Sklare: *The Technician Writes: A Guide to Basic Technical Writing*

©1971 by Boyd & Fraser Publishing Company.
All rights reserved.
No part of this work may be reproduced or used in any form
or by any means—graphic, electronic, or mechanical, including
photo-copying, recording, taping, or information and
retrieval systems—without written permission from the publisher.
Manufactured in the United States of America.

Library of Congress Catalog Card Number: 72-141220

ISBN: 0-87835-013

2 3 ● 3

For
I.S.S.

Preface

This is an English book on writing for the technical student. He may be in a two or four-year general or special college, a community college, or a technical school or institute. The book tries to show this future technician how to write when on the job. Emphasis is upon the basic and practical. Having mastered the text, the new technician will require on-the-job writing experience to round out his training. But the book is designed to prepare the technical student adequately for those initial writing demands he will normally face in industry.

Technical and special education strips English training to the bare bone. A technical man, unfortunately, may not always have the benefit of the full-year freshman writing course taken by those not in his specialized program. Our book is planned to satisfy the special needs of the beginning student in technical English. The future technician, responding to economic and industrial realities, is obliged to concentrate his writing training on practical, useful matters. Our book is for him.

There are four parts. Part I is comprised of ten chapters on the technical report. Part II devotes four chapters to other technical communications—letters, memoranda, articles, and abstracts. In Part III four chapters deal with some of the refinements of technical writing—functional English, sentences, paragraphs, and matters of grammar. There are two reference guides in Part IV, one to good usage and the other to grammatical terms.

While the text, therefore, is neither a freshman English handbook nor a rhetoric nor a book of readings, it attempts to incorporate and adapt to the needs of the technical student some of the usual features of each of these standard tools. The first part, for example, contains ample material on technical library research. The traditional expository modes are shown as instruments of technical writing. Handbook matters of use and grammar are highlighted in the last two parts. Relevant readings from the technical and scientific literature provide enriching supplementary views for several of the key chapters throughout.

While we hope the book pleases professors, reviewers, and librarians, real approval for it must come from the aspiring technician who is struggling to communicate with his world. If he is helped on his way, the book is justified.

Arnold B. Sklare

ACKNOWLEDGEMENTS

The Faculty Research Committee of Southampton College of Long Island University provided a grant-in-aid for the writing of this book.

Members of the Southampton College Library staff were unfailingly helpful in locating and obtaining materials.

Many authors, publishers, companies, and professional organizations generously granted permission to reproduce materials belonging to them.

Mrs. Sue Grant faithfully prepared the manuscript.

Mr. Arthur H. Weisbach of Boyd & Fraser Publishing Company provided sustaining advice and encouragement throughout.

Grateful thanks and acknowledgement are due to all.

Table of Contents

Preface ... iii
Acknowledgements ... iv

PART I: THE TECHNICIAN WRITES A REPORT

1. What Is a Technician? ... 3

2. The Technical Report ... 13
   A SUPPLEMENTARY VIEW: "Design That Report!" *by James W. Souther* ... 23

3. The Preliminary Plan ... 27
   A SUPPLEMENTARY VIEW: "How To Start To Write a Report" *by J. Packard Laird* ... 32

4. Methods of Gathering Information ... 37

5. Preliminary Bibliography ... 50

6. The Job Technical Writing Does ... 56
   A SUPPLEMENTARY VIEW: "The Method of Scientific Investigation" *by Thomas Henry Huxley* ... 70

7. The Job Technical Writing Does (Cont.) ... 73

8. Parts of the Report ... 88
   A SUPPLEMENTARY VIEW: "Suggestions for the Preparation of Technical Papers" *by Robert T. Hamlett* ... 115

9. Presenting the Completed Report ... 123

10. Visual Aids ... 139

PART II: THE TECHNICIAN WRITES LETTERS, MEMOS, ARTICLES, ABSTRACTS

11. The Business Letter — 159

12. The Memorandum — 183

13. The Technical Article — 191

14. The Abstract — 209

    A SUPPLEMENTARY VIEW: "The Writing of Abstracts" *by Christian K. Arnold* — 215

PART III: THE TECHNICIAN WRITES FUNCTIONAL ENGLISH

15. Functional English — 223

    A SUPPLEMENTARY VIEW: "What Can the Technical Writer of the Past Teach the Technical Writer of Today?" *by Walter James Miller* — 227

16. Functional Sentences — 246

    A SUPPLEMENTARY VIEW: "The Seven Sins of Technical Writing," *by Morris Freedman* — 253

17. Functional Grammar — 262

PART IV:

The Technicians' Guide to Good Usage — 273
The Technicians' Guide to Grammatical Terms — 295

Part I
THE TECHNICIAN WRITES A REPORT

Chapter 1
WHAT IS A TECHNICIAN?

A textbook on basic technical writing might very logically begin by answering the question: What is a technician? The simplest, clearest answer is in the brochure entitled "25 Technical Careers You Can Learn in 2 Years or Less," produced and distributed through a cooperative effort of the U.S. Office of Education and the National Industrial Conference Board:

> Roughly speaking, technicians are people who work directly with scientists, engineers and other professionals in every field of science and technology.
>
> Technicians' duties vary greatly, depending on their field of specialization. But in general, the scientist or engineer does the theoretical work. And the technician translates theory into *action*.
>
> The best-known technicians right now are probably the ones who work with computers. But there are many other kinds.
>
> Look below. You'll find a list of the types of technicians now in the most demand—along with a brief description of the kind of work they do.
>
> AERONAUTICAL AND AEROSPACE—Work on design of space vehicles, missiles, supersonic transports. Help solve air traffic control problems. Help explore space. The aeronautical and aerospace field is growing so quickly, these are just a few examples of the work this technician does.
>
> AIR CONDITIONING AND REFRIGERATION—Help in the design of future astrodomes, spaceships, sea laboratories, ultramodern homes and cities under domes—the air-conditioning, refrigerating and heating systems of the future.

AGRICULTURAL—Work on the scientific production and processing of food and other things that grow. Act as consultant on farm machinery, agricultural chemicals and production techniques.

ARCHITECTURAL AND CONSTRUCTION—Work on projects to rebuild our cities. Develop new building techniques, and new materials for building. Through city planning, help with the sociological problems that plague our inner cities.

AUTOMOTIVE—Assist in the design of new traffic control systems. Help in the planning of tomorrow's cars, especially smog control devices, automatic guidance systems, and new safety features. Work on the problems of mass transport.

CHEMICAL—Work in new fields of chemistry, especially biochemistry, chemical engineering. Help develop new materials from chemicals, especially new plastics, new foods, new fertilizers, new anti-pollution agents.

CIVIL ENGINEERING—Work with computers to solve structural problems in constructing new buildings. Work on automatic highways. Help rebuild our cities and our highway systems. Work on unconquered environments (like the jungle, the ocean) to make them livable for man.

COMMERCIAL PILOT—Pilot airplanes and helicopters. Act as cabin crews in commercial aircraft. (These careers call for experience after your 2-year educational program.)

ELECTRICAL—Work with new electrical devices—like portable power systems for use in remote areas, fuel cells for use in spaceships, sea labs. Help design electrical systems for tomorrow's homes, factories, etc.

ELECTRONIC—Work in new electronic fields like miniaturization, solid-state devices. Work along with scientists in new bioengineering systems. Work on fourth-generation computers, teaching machines, etc.

ELECTROMECHANICAL—Help design new information systems, new computers. Work on artificial hearts, other human organs in the new field of biomedical technology. Work on automated production equipment.

ELECTRONIC DATA PROCESSING—Process and analyze business and scientific data using new generation computers. Develop new systems analysis to solve storage and retrieval problems. Help develop new teaching machines.

FIRE PROTECTION—Develop new fail-safe systems for supersonic transports, sea labs and other artificial environments to make them safe for human habitation.

FORESTRY—Help care for, protect and harvest forests. Develop and conserve wildlife and recreational resources.

HEALTH SERVICE—Work with medical teams as assistant or nurse on the new frontiers of medicine. Work on bioengineering techniques to save and prolong life. Work in dentistry and oral surgery.

INDUSTRIAL PRODUCTION—Help design new production methods—particularly automated systems—and new materials, machinery, and control systems to make industry even more productive.

INSTRUMENTATION—Work on the instruments that have brought about space exploration, new medical techniques, automation, pollution control and other modern miracles.

MARINE LIFE AND OCEAN FISHING—Develop new procedures for harvesting food from the ocean. Help discover new minerals beneath the sea. Work on conserving the ocean's natural resources.

MECHANICAL DESIGN—Work on producing new kinds of machines for tomorrow's manufacturing plants, hospitals, homes.

METALLURGICAL—Help develop and produce new "miracle" metals and metal alloys for use in construction, machinery, medicine, etc.

NUCLEAR AND RADIOLOGICAL—Help research, develop and produce nuclear devices and atomic power plants. Use radio isotopes in industrial and health fields.

OCEANOGRAPHY—Explore the ocean's chemistry, geography and mineral resources. Develop ways to use them. Develop manned underwater sea labs.

OFFICE SPECIALIST—Accounting, financial control and management. Scientific, legal, medical or engineering secretaries.

POLICE SCIENCE—Work on new, scientific methods to detect and prevent crime.

SANITATION AND ENVIRONMENTAL CONTROL—Help improve man's environment and protect natural resources by scientific means. Help prevent or control air and water pollution. Inspect and prevent contamination of food. Control waste disposal.

WHAT IS A TECHNICAL WRITER?

While we now know what a technician is and the kinds of work each technician may do, the question of what a technical writer is still remains to be answered. In an effort to define a technical writer, Fordham University researched and prepared "A Report of a Study to Determine the Duties and Responsibilities Called for Under the Job Entitled Technical Writer." Based upon information obtained from leaders in the field of technical writing, the Fordham study compiled this job description of the technical writer:

> The technical writer is one who writes instructive or descriptive material on technical or scientific subjects, interpreting and creating an acceptable presentation of the facts or the ideas and theories of others for a given audience.
>
> The work includes the following:
>
> 1. Performs research necessary to obtain complete understanding of the scope of the proposed publication and to gain a thorough technical knowledge of the subject.
>
>   Receives a verbal or written work order for the desired publication, together with instructions on its general purpose, and any available basic reference material, such as specifications, proposals, correspondence, engineering reports, drawings, photos, similar publications, and supervisory or sales memoranda and notes.
>
>   Studies the supplied reference material to acquire background information on the project and to ascertain policy governing content, presentation, on the quality level. May consult with engineers, other technical personnel, the publications supervisor or sales personnel to clarify technical or other details of the writing project.
>
>   Analyzes information on hand to determine whether additional research is required or whether the supplied material is sufficient and can be adapted to the publication requirement.
>
>   If additonal research is required, determines the most logical sources and the best method for obtaining the required information. Performs the necessary research, may make field trips to libraries, government agencies, manufacturers, educational institu-

tions, technical societies, etc. May confer with customer's technical staff through established lines of liaison and may observe, study, or operate the actual equipment, object, or process.

Makes suitable notes to ensure proper correlation and retention of the information obtained.

2. Organizes the proposed manuscript to provide an orderly plan for the preparation of the required text material. Prepares a general outline; breaks the subject material into major topics, considering:

   a. The general purpose of the manuscript (catalog, magazine article, engineering report, equipment operation or maintenance manual, etc.).
   b. The specific application (formal training, guide for field operations, promotion, general information, etc.).
   c. The knowledge and skill level of the user.
   d. The complexity of the subject.

   Arranges these major topics in logical order. Determines the logical sub-topics to be discussed or treated under each major topic and arranges these in proper sequence. Classifies and indexes the reference material in accordance with the general outline.

   Prepares a detailed outline; analyzes the reference material for each topic and develops and expands ideas into further sub-topics and groupings and arranges them to achieve continuity and best subject coverage.

   Repeats this procedure for each topic, developing the outline for smaller and smaller portions of the manuscript, to the logical ultimate.

3. Prepares a draft of the manuscript in accordance with the detailed outline. Writes the text, drawing upon his developed knowledge of the subject and desired scope, and using his communications skills to create an acceptable presentation of the technical data for the given audience. May conduct additional research to validate or clarify portions of the technical data. Uses a style and format for the writing set forth in applicable specifications or may select or develop a style or format best suited to the presentation.

   Defines new and unusual terms.

   Determines the illustrations required to supplement the written material and selects the most suitable type of illustration, such as a photograph, line drawing, rendering, etc. Prepares sketches or preliminary layouts of line drawings and renderings and specifies the requirements for photographs.

May supervise the photography. Assigns nomenclature to photographs by marking on overlays or other method.

Requests the preparation of preliminary or final art from the art department and provides additional oral or written instructions as required.

Maintains written control and record of changes in cross references, figure references, tables and the like during the development of text and illustrations to ensure accuracy of these details in the final manuscript. Routes the final manuscript through established channels to obtain technical editor or customer approval.

4. Revises and rewrites text to meet technical editor's and/or customer's review requirements.

Receives the draft of the proposed publication after technical editor or customer review. Studies the corrections, comments, criticisms, or suggestions made to determine the specific revision requirements and their effect on other portions of the text.

Rewrites affected portions of the text and requests new or revised illustrations as required. Checks very closely to assure that all references and notes in other portions of the text conform to the received portion and makes any required changes or corrections.

Reviews the new or revised illustrations to ensure accuracy and conformance with required changes.

Routes the revised text and illustrations for final approval. May obtain and present factual data as a basis for not accepting changes requested by the editor or customer.

The technical writer is responsible for the development and presentation of text and illustrations for technical publications which may cost thousands of dollars. He is responsible for completing his work within the budgeted hours under maximum general supervision; for meeting acceptance standards and delivery schedules for the completed manuscript; for technical accuracy of work performed by illustrators, typists, and others engaged in producing the manuscript; for determining the necessity for liaison and research.

In addition, he must be able to interpret technical and scientific data, such as blueprints, diagrams, charts, engineering reports, and specifications for material equipment, publications; he must know research methods and techniques and be able to plan and organize manuscript in accordance with the requirements of specified media.

The writer must know his tools which include a comprehensive knowledge of good grammar and punctuation. He must have the ability to write clear and concise descriptive and instructional material, and understand illustration techniques and publication production methods and practices. He needs a general knowledge of the basic sciences and specialized training or experience in the technical area in which he is writing, i.e., aeronautics, agriculture, electronics, chemistry, mechanics, etc.

In addition to all this above, the technical writer needs the ability to discriminate between essential and non-essential data from the reader's viewpoint and thereby determine sufficiency of content. He must keep abreast of current trends and techniques for written communication and their particular application to his specific work and be able to organize major publication projects so as to determine suitable work assignments for assistant writers. His writing should be readily adaptable to writing styles and formal requirements. As a skilled writer he must carry out several projects concurrently and arrive at decisions as to the relative merits of different methods with respect to their effect on time, cost, and acceptance standards for the end product.

This represents a composite picture of the "perfect" technical writer. Perhaps there is none in reality but it does provide one with specific job guide-lines.

TYPES OF TECHNICAL WRITING

It is understood that probably only a few readers of this text are preparing for the career of a professional technical writer. The job description given above was presented to acquaint the beginning technologist-writer with a broad picture, since various responsibilities falling to the professional technical writer will necessarily touch on the experience of any technician required to write as part of his job. The present text, however, limits itself to those phases of technical writing most useful to the person employed in business or industry primarily as a technician and not as a writer.

Though writing is not the main function of such an individual, unless he is able to master basic technical writing, his usefulness in business and the opportunities available to him are fewer than for one able to write well. Chiefly, the technologist in science and industry is required to write reports. Accordingly, the present text will concern itself largely with the technical report. But since the technologist-writer may frequently need to write business and technical letters, technical memoranda, and abstracts as part of his function in industry, these forms will also be covered. In addition, since a technician who reaches proficiency in his field may prepare a technical article for publication, article writing will be considered as well.

There are also thousands of technologists employed *primarily* as technical writers, and they are engaged in many different assignments. Beyond the forms of writing mentioned above—those employed by the technician in business or industry—there is an increasing variety of technical writing presently being done. But all varieties of technical writing, professional or other, require the writer to gather data, organize the material, write the text, and prepare the final publication that will communicate scientific information to readers.

The magazine article on scientific and technical subjects is perhaps the form of technical writing best known and most in demand. Such articles are published for varying audiences. Engineering or scientific societies publish their professional journals. Technical magazines are produced by commercial publishers. Trade magazines exist in most fields as well. Each of these appeals to different reading groups.

Another form of magazine is the house organ, a journal for distribution inside a large company to employees and outside to potential customers. Such magazines sometimes are widely circulated to schools, libraries, and foundations as a means of enhancing the reputation of the company.

The professional technical writer, apart from the writing of magazine articles, may be involved in preparing pamphlets or brochures for publication. These communications are directed not only to outside customers but to management and other units within the company to keep them acquainted with ongoing developments. Such pamphlets and brochures are important instruments in selling the company's products and are therefore written with skill and care.

Technical books represent perhaps the most formal type of technical writing carried out by the professional. They occupy a position of high priority in the publishing industry. As with magazine articles, they appeal to a wide range of audiences. Some may be written to a limited readership of, for example, scientific theoreticians in a particular field. Others may be directed to technicians, while yet another group of technical books may be broadly directed to the general reader who has merely a casual interest in the subject at hand.

Instruction books or manuals are yet another type of writing assignment which might fall to the full-time technical writer in industry or government. Most manufactured equipment requires an instruction book, particularly in the military. Specifications must be exact and instructions clear, since many potential readers of the manual may not be technically skilled. Complicated mechanical and electronics systems must be explained to buyers, maintenance personnel, and dealers in business and industry. Preparation of technical instruction manuals requires both writing skill and the ability to grasp challenging technological data and principles.

Yet another type of technical writing is carried out for advertising, sales promotion, and publicity. Here the advantages and special features of a technical product are explained in detail to the potential buyer or user who may be a physician, or an engineer, or a purchasing agent in government or industry. The professional technical writer here not only must know his own product and his competitor's product, but must possess, in addition, knowledge of selling techniques.

The student may begin to see the range and scope of opportunities and challenges offered by technical writing. Technical writers write about almost every product and service in industry. The form of their writing may vary from reports for management to books for the masses. The satisfactions and rewards to be gained by the technician who learns how to write well should be a strong motivation to the student beginning his study of technical writing. Hopefully, among those who are presently technician-writers are some who will in the years ahead become writer-technicians and help fill the need now urgently felt for these valued specialists.

WRITING, TECHNOLOGY, AND THE TECHNICIAN

Technical writing, then, is the communication of useful information. Its degree of usefulness depends upon the technical writer's skill. Fewer people will share the benefits of technology if the proficient technician is unskillful in written communication. Technical writing is public writing in this sense of sharing useful information.

If the technically trained person has not been a strong student of English in primary and secondary school, this need not hinder his mastery of technical writing in college and in his technological field. A technician learns to master his subject in broad outline and in minutest detail. Accordingly, he is well disposed by training and cast of mind to become a competent technical writer. The method of logical procedure and the ability to see relationships between parts that make for successful technology are also required for the presentation of a successful technical report.

While the technician often does his daily work as part of a group, as a technical writer he usually works alone. Technical writing, like all writing, is lonely work. The most trustworthy support the individual technical writer can have is his own grasp of technical-writing principles. After practice in translating his technology into standard forms of written presentation, the technician will feel less isolated in his reporting. He will return to his field or shop or laboratory work not only with a fuller technical knowledge but with a richer sense of participation resulting from his technical-writing experiences.

This reciprocal relation between technology and technical writing should be acknowledged. Everybody wants to be a technician, but few are willing to submit themselves to the discipline necessary to become a technical writer. Technical writing spurs the individual technician's personal professional growth and adds generally to the advancement of technology.

Chapter 2
THE TECHNICAL REPORT

OBJECTIVE

You are required to write a technical report. Where do you begin? You begin, first, by determining the objective of the report and, second, by drafting a preliminary plan which appears able to fulfill the objective you have determined.

HOW TO DETERMINE THE OBJECTIVE OF A TECHNICAL REPORT: THE READER

The reader of a technical report is the most important person the writer must consider.

Who is the reader?
What does he want to know?
What does the reader already know?

The technical writer must answer these questions as well as he can, as early as possible. The preliminary plan for the report—the terms of the information to be presented and the form of the presentation—will largely depend on the reader.

As a student and beginning technical writer it will not be difficult for you to know who your reader is—probably your writing instructor or a professor in a technical course! But once you leave school, you will quickly see how widely readers of technical reports vary in age, education, and experience. They will differ not only in their jobs and in their training, but also in the uses they make of your report.

For you as a beginner, it is best to assume that your readers are your peers—persons generally of your own age, education, experience, and interests. This will be an acceptable learning situation only if you recognize it as fairly ideal and therefore artificial.

And so you know who will read your first technical reports. But it is obvious that even among your own general age and education group there will be wide individual variations. Can the reader of your report possess just the same skill and training in mathematics as you? In physics? Will his experience be greater or less than yours, for example, in machine–shop practices? Even in reporting to your professors, or to your peers, you will be unable to determine the objective of your report effectively without considering these matters.

The objective of a technical report must be seen in terms of the needs of people: the readers.

> *Report Objective:* To explain *to shop supervisors* in the western district the method of assembling and installing the condensor-component wiring for the AL 34 amplifier.

> *Report Objective:* To report *to the field workers of the department of environmental sanitation* the results of spraying alfalfa with chemical substance VERPLOX during a 6-week trial.

As you proceed from the formulation of your report objective to a preliminary plan to fulfill it, you will see again that the readers of your report will necessarily determine not only the level of technical content you include, but the organization, data, and style of presentation as well.

Once you determine who your reader is, making up your mind about what he will want to know—and what he already knows—about the subject of your report will not be so difficult. In the daily work of business and industry, a report does not exist in a vacuum. The very need for reporting springs from the high degree of organization and specialization in technology today. While it is likely that most of your technical reporting will be directed to other engineering and technical personnel, your readers will sometimes be unskilled or semiskilled workers, staff personnel, or managerial personnel. What readers from any of these groups already know, or may wish to learn about the specific subject of your technical report, will vary considerably.

You, for example, are a knitting machine technician, or an aircraft design technician. You are preparing a report on safety regulations for the operation of simple power equipment. You can safely assume no prior information on the subject if your report readers are unskilled workers. Accordingly, they will wish to know in complete detail every precaution they must exercise in the daily operating of their phase of the equipment. A report on the same subject for the technical employees in the same plant would concern itself with different matters, since these workers already know a good deal about the equipment. They will wish to know those precautions for safety relating, for example, to installing, repairing, and maintaining the hardware rather than operating a certain part of it on a daily basis.

Who the reader is, what he wants to know, and what he already knows are in these ways important in deciding the objective of the report in terms of people.

*FIVE LEVELS OF TECHNICAL WRITING**

Since it is of prime importance to anyone engaged in technical writing to know the proper level at which the material should be written, attempts have been made to classify audiences into levels according to the assumed background of knowledge of average readers in each level. General rules have been developed for writing to each level.

Five levels of technical writing which have been established are:
1. Operator's or non-technician's level.
2. Field or technician's level.
3. Depot, advanced technician's or junior engineer's level.
4. Engineer's level.
5. Advanced engineer's or scientific level.

---

*Adapted from J. Racker, "Selecting and Writing to Proper Level," *IRE Trans. Engineering Writing and Speech*, Vol EWS-2, No. 1, January, 1959.

Writing rules can be established for each of these five levels. Students of basic technical writing, however, will be mainly, if not exclusively, concerned only with the first three levels.

1. **Operator's or non-technician's level.** The average reader of a technical report at this level generally has no specialized training in the subject matter of the text. He does, nonetheless, often possess a fragmented technical background and a broad knowledge of mechanics, but it is wiser to assume that these are minimal.

Several typical examples of situations may be helpful. The writer wishes to describe the operation or maintenance of equipment to a trainee-operator with no previous experience with such equipment. Or, the writer wishes to instruct a medical doctor on the operation of an electronic diagnostic device. From the standpoint of his knowledge of electronic equipment, the doctor is considered a non-technical operator. Or, the writer wishes to present the overall advantages and disadvantages of a piece of equipment or system of operation to corporate management. The management reader is not particularly interested in technical details. Mainly, the non-technical aspects of overall costs and procedures would be of interest to him. In this light, the management reader is on the non-technical level.

Guidelines for Writing to Level 1.

Provide all Information.

Do not place unfair demands on the reader by expecting him to supply technical information necessary to understanding. If sufficient information is lacking, the reader will feel justified in shunting the report aside.

Support the Text with Clear, Graphic Illustrations.

Pictorials and graphics will help emphasize and clarify the text. Make the illustrations simple and as close as possible to the physical appearance of the equipment under discussion. Avoid using schematic or symbolic representations. (Cartoons have come into use in recent years in an attempt to add clarity and interest to technical material written to the non-technical reader. The present author finds these cartoons generally ineffective, but each student must make his judgment about their use.)

Emphasize Interlocking Step-by-Step Procedures.

Present explanations in distinct, short steps. In placing the steps on the page, exploit all the devices of outlining, indentation, typography, numbering, and labeling to make them clear. Arrows, flashes, brackets, etc., are often useful to connect or subdivide procedural instructions.

Avoid Reference to Other Instruction Procedures or Manuals the Reader May Not Know or Have.

Give the reader all the steps and instructions necessary to carry out or understand the operation without having to seek elsewhere to fill in information. Assume that the reader is not richly familiar with standard operating procedures for any separate phase of the process. Briefly and clearly spell everything out.

2. **Field or technician's level.** The typical reader of a report at this level is considered to have special training and experience on a particular class of equipment or system of operation. He is assumed to possess basic tools and test equipment related to his technology. In general, such a reader is thought of as having been trained to maintain equipment in the field.

A reader at this level might be a home-appliance repairman who services equipment in the home. His special competence might be in either automatic washing machines or refrigerators. Or, the writer may be addressing an automotive mechanic at this level. Like the appliance repairman, the mechanic is trained to repair common breakdowns, make some adjustments, remove major assemblies, and perform specified routine maintenance procedures. Or, the reader in this classification could be any one of a large number of technicians trained to service equipment in the field by carrying out prescribed preventive and corrective lubrication, adjustment, and replacement of specified moving parts.

Guidelines for Writing to Level 2.

Indicate, Whenever Possible, Testing Equipment Available in the Field.

In most commercial operations, the kinds of field test equipment normally available to the technician are generally known by practice and experience. Maintenance work requiring test equipment or parts not available in the field is appropriately considered depot maintenance. There is no need, in writing to this level of communication, to provide detailed operating instructions; it may be safely assumed that the technician will possess this knowledge. When special procedures, however, are to be followed in the field, the writer is obliged to supply full information on the setup and operation.

Provide a General Theory of Operation of the Equipment so that the Field Technician Can Better Deal with Specific Problems He May Encounter.

No report can offer maintenance information to account for every possible difficulty the field man may experience with the equipment. With a theory-of-operation section, however, general procedures can be outlined and common causes of failure suggested.

It is not necessary to repeat basic theory or to describe details of theory that will not aid in the maintenance of specific equipment at the field level. To support his explanation to the field technician, the writer may provide simplified schematic drawings.

Employ Graphs or Visuals in Place of Equations, Whenever Possible.

While the field technician may understand simple mathematics, formulas should be supplemented by illustrations. If possible, use of mathematics to explain theory or to provide maintenance instructions should be kept minimal.

Refer to Standard Instruction Procedures and Manuals; It Is Unnecessary to Repeat Instructions.

The working field technologist may be reasonably expected to have available to him procedures and manuals commonly referred to by those similarly skilled.

3. Depot, advanced technician's or junior engineer's level. The representative reader at this level of technical writing may be assumed to have rich background information on the subject; he has had a good deal of experience or training. Accordingly, he can fully understand the theory of operation of the equipment. He knows the appropriate tools or test instruments needed to maintain the machine when it is sent to the depot, shop, or factory for work.

Perhaps such a reader is the experienced troubleshooter who repairs and adjusts equipment that cannot be handled by the field man. Or, he may be the advanced technician capable of overhauling and rebuilding complicated hardware. Readers at this level of communication are competent at building and testing developmental models of equipment—that is, they can design test setups and make minor design modifications.

Guidelines for Writing to Level 3.

Eliminate Explanation of Standard Test Equipment or Procedures.

Personnel at this level are expected to know, to be able to procure, and to be competent to operate available test equipment. Such readers may modify available test equipment, to make it equivalent to equipment specified, in carrying out diagnostic procedures.

Describe Special Test Equipment or Unique Procedures.

If there are unique operating instructions or special test equipment, these should be explained.

Explain Any Special Disassembly and Assembly Techniques; Provide Supplementary Diagrams.

While reports at this level sometimes include explanation of complete disassembly and assembly procedures, they always include special techniques. To supplement the text, illustrative diagrams show the disassembled equipment and identify each part. The order of disassembly may be shown by index number of each part. Such diagrams are useful for reassembly.

Include the Complete Theory of Operation of the Equipment.

The theory of operation should cover all aspects of the equipment and include complete diagrams to permit servicing all items in the equipment.

Employ Graphs and Visuals When Possible.

Refer to Standard Instruction Procedures and Manuals; It Is Unnecessary to Repeat Instructions Available to the Technician Elsewhere.

Guidelines for writing to level 4.

Provide Basic Background Information the Reader Requires.

If basic introductory background material is not provided, the reader should be told what information is required and where it may be obtained.

Maintain Objectivity Throughout.

Emphasize demonstrable facts and characteristics.

Provide Complete Information.

It is essential to provide full facts and steps leading to a conclusion. The writer must not assume the reader shares the personal approximations and assumptions that lead to a particular conclusion.

Use Mathematics Freely, as Necessary.

Mathematics should be included when necessary to achieve the objective of the writing.

Define All Terms in Equations.

Employ the Same Units Throughout the Paper.

The reader should know when the writer shifts from inches to centimeters, or from grams to ounces. If it is necessary to do this, a conversion factor should be specified.

Guidelines for writing to level 5.

Cite Sources Where Basic Background Information May Be Obtained.

Freely Employ Higher Mathematics, Where Necessary.

Provide Complete Information Leading to Conclusions.

Define All Terms in Equations.

Employ the Same Units Throughout the Paper.

HOW TO DETERMINE THE OBJECTIVE OF A TECHNICAL REPORT: THE PURPOSE

The purpose of a technical report is determined by the use to which the reader will put it. While you as writer are framing the objective of your report and drafting a preliminary plan for it, you will wish to know whether the reader will use the report for information, decision-making, or record-keeping. In actual practice the three broad areas of report usefulness may be overlapping.

Readers seeking information may be either outside or inside one's own company. Reports written in response to requests for information from customers about the use of company products are information reports. Reports used inside the company to record progress on a project; or to record the results of an investigation or inspection; or to give data on materials or products tested; or to record a failure of equipment or an accident—all of these and many more are reports the writer prepares with the basic purpose of transmitting information to the reader.

Decision-making or action reports likewise provide information, but the information is used to produce an action. Reports which propose a course of action to the reader are action reports; standards and specifications reports which control methods in construction fall into this category. Written requests or authorizations for jobs to be done; instructional reports which transmit information about the job; reports which make a bid and provide an estimate—these are technical reports for decision-making or action by the reader.

Record-keeping reports help provide running accounts of a project. Pre-printed forms are often used for these work reports or service reports which record data and keep a week by week history of the progress and costs of a particular undertaking.

In sum, the beginning writer recognizes that the broad and all-inclusive purpose of technical writing is to communicate useful information clearly. The writer's job ultimately reduces itself to a specific subject directed to a specific audience for a specific purpose. The writer obtains and organizes his facts, and he presents them in the most clear and simple fashion possible to achieve his objective most effectively.

a supplementary view...

DESIGN THAT REPORT! *By James W. Souther*

From Journal of Chemical Education, *Vol. 28, October, 1951. Reproduced with permission.*

Too many courses in technical writing still reveal their being born of traditional composition courses, while too few possess any close relationship to actual writing situations found in industry. Contact with professional technical report-writing problems has indicated two main difficulties: first, the failure on the part of the writers to *design* their reports to fit a specific situation, purpose, and audience, and second, a lack of knowledge of the *process* involved in writing. This is not a criticism of the technically trained men, but rather of the education given them. To ask a person to design a gear assembly for a particular mechanism without providing both the context of the problem and a training in the required skills would be considered foolhardy; yet that is precisely what is being done in a great many report-writing courses.

Experience teaches that too many writers fail to analyze the report-writing problem which faces them. Instead of first considering how the conditions of the assignment, the purpose of the report, and the users of the report should influence their work, they want to start writing long before they know what the report is to be, what it is to do, what it is to say, or who is to use it. Furthermore, competent technical men often fail to consider how the time given them to write a specific report affects what should be included and how long the report should be. If what they feel has to be included is more than they have time to do, they work overtime and complain of the load they have been given. Of course, many times the assignment is given neither clearly nor concisely enough; also too much may be expected in too short a time; however, more often than not, the writer has failed to recognize the influence of the time element in the assignment which produces extra work.

Likewise they often fail to realize how they, as writers, should be affected by the aid which is available to them and by the amount, kind, and sources of accessible information. One man was faced with the task of periodically compiling a master report from those submitted to him. Although he was an excellent engineer and succeeded in filing a good report, he was doing a great deal more work than was required because he completely rewrote all the material submitted to him. The aid which he got was not being fully utilized, for his work would have been much easier had he sat down with his assistants and provided them with the over-all pattern of the master report, making sure that each understood his part in the formation of the whole. He could then merely have compiled the report, adding whatever was necessary, rather than have rewritten it.

In addition to the elements of time, information, and aid, two other important conditions must always be taken into consideration: the specific requirements imposed upon the writer and adherence to company policy. If the request for the report is accompanied by "include the causes of production line stoppages last month," the content of the report and the performance of the writer are affected by that order. Moreover, general company policies and style-sheet requirements are always in the background and must be considered.

Obviously a report which is to be used for later reference requires coverage, data, and detail different from that in a weekly progress report; yet many experienced chemists do not realize that these differences exist because they have not been made conscious of the requirements which the purpose of the report places upon them. For example, at one concern a large staff of chemists was employed in the experimental laboratories, and part of their duties was to record their experiments and findings so that these could be used later, whenever the need arose. These reports proved, too often, useless for such purposes, for not only could the results not be clearly understood but, sometimes, the experiments could not even be repeated. An analysis of the situation indicated that the chemists did not realize that reference use of the report placed special requirements on them as report writers. At the time of writing, the experiments were fresh in mind; the writers were comfortably in the context of the situation, and little thought was given to what would be required by the reader in order to understand the account when he came to it "cold," sometimes years later.

An example from another concern also reveals this same failure. One of the men presented a long report which was to be used for constant reference by non-technical personnel; yet the work was rendered almost useless because of the general "layout." After a table of contents, thumbtabs for the main sections, and an index were added, the report became both usable and valuable. These devices are quite common. The writer knew of them but he neglected to use them because he had not *designed* his report to serve its purpose.

More and more, writers in industry are becoming aware of their readers' interests. They are placing conclusions, summaries, and recommendations at the beginning of the report because the administrators are most interested in such material. The more detailed material is placed toward the end of the report, for often only the specialist assigned the job of checking these details will read it. The more widespread use of such devices as statements of purpose and background, abstracts, summaries, and conclusions at the beginning of a report is ample proof of the writer's growing awareness of the reader.

In addition the writer should consider his own background in determining what is required in the report. If possible he should answer the question "Why was *I* asked to write this report?" In many cases the answer will provide an indication of what should be in the report or what general approach should be used. For instance, in one concern two men with quite different backgrounds work in the same department. One has worked his way up through years of practical experience in several concerns, and the other is a college graduate with a few years' experience in this particular concern. If a writing assignment were given to the older man, he could be quite sure that the "boss" wanted the views of a practical man with a wide background of experience, and he should emphasize his material accordingly. If an over-all picture were needed, the younger man would be asked to write the report. In both cases the effectiveness of the writers would be increased by an analysis of what their backgrounds could add to the writing requested of them.

Recent classes in report writing offered to engineers and chemists actively engaged in industry revealed, in almost all cases, a lack of understanding of the activity of writing or of the process involved. This is not surprising, for the emphasis in their training was on the finished product, not on the activity which produced it. As a result, when faced with a writing problem they thought in terms of the finished product. They tried to write the report without planning it first, like a carpenter trying to build a house without blueprints. Only confusion and difficulty could result.

Once they started with the analysis of the writing situation—the assignment conditions, the purpose of the report, the audience, and their own background—they found writing much easier. When they had these restrictions well in mind, they found that they could focus on the topic of the report more effectively, and, of course, they could more easily tell what material was necessary because they knew what was expected. The gathering, evaluating, and selecting of material became a relatively simple task because the analysis of the situation had provided them with the necessary standards from which they could work.

When the selection was completed, however, they found themselves facing a completely different task, one of constructing rather than analyzing. The arrangement of material into logical patterns, into main and subordinate classifications, into meaningful relationships, became easier because these considerations depend to a large extent on the writer's purpose and intent, on what he wants to do. Once they knew what they were trying to accomplish, they could, by keeping their purpose in front of them, organize the material in meaningful stages and indicate the relationships between the stages. They began to realize that these relationships were what gave their work real meaning and value. More important, they found that the prospect of actually writing the report no longer seemed overwhelming. They had reaped the fruits of designing their reports: they knew where they were going and what they wanted to say.

Of course, they had not reached the end of the writing process; they were still faced with the writing and rewriting, but their particular difficulties had, in the main been overcome after they had gone through the planning process. The rest seemed, in most cases, to take care of itself. Surprisingly enough, many of the grammatical and stylistic problems eliminated themselves, for they seemed to be the result of confusion and indirection rather than of a basic lack of knowledge.

These experiences with acutal industrial writing problems suggest that we who teach technical writing need to adjust our courses so that they more closely resemble the professional situation. They further suggest that we need to train our students to analyze the writing problems and to provide them with some understanding of the process involved in writing. It seems also that we need to shift our emphasis from the exactness of the finished product to the activity of writing. We need to help the students reach the finished product; moreover, we should criticize the work from the point of view of how well it achieves its purpose and fits the situation rather than merely correcting it in a grammatical vacuum.

One thing is certain: a *realistic approach* to technical writing is more likely to be found through actual contact with industrial writing problems than by simply continuing the traditional composition approach so often used. This of course places a tremendous burden on the shoulders of the teacher because it means, first, that every writing assignment must be given a context—that is, a specific situation and problem must be created for each report the student is to write—and second, that each student must be trained in the skills of the writing process. This is the realistic approach, for it is the one that best simulates professional conditions.

Chapter 3
THE PRELIMINARY PLAN

Report writing moves though logical sequences from start to finish. At no point is this logical continuity clearer than in the movement from the statement of the objective to the preparation of a preliminary plan to fulfill it.

Any preliminary plan will attempt to lay out for the technician a check-list of the work to be performed. Such a plan will at first be prepared in broad outline, with details supplied later. The writer attempts to set down in logical order the various steps and procedures he will probably need to follow in achieving the objective of his report. This preliminary plan may be thought of as a guide not only to help the technician perform the work necessary for the report but to provide a structure for the report itself.

A portion of this plan will necessarily concern itself with materials needed and methods to be followed. As to materials, is all equipment available and are necessary supplies on hand? What are the complete resources one will need to work with as these are forecast at the beginning of the project? And what methods and procedures will the technician follow in working his materials in a certain order so that he may best achieve his particular objective? Which steps must be carried out first? How can the plan for work be organized so that it will eventually lead to a report of maximal usefulness for a particular reader?

In its rawest form the preliminary plan may be no more than a listing of what the writer must do with what things to achieve the objective set. Once these elements are down on paper, some time can be spent in bringing them into what appears to be the most logical relationships. In setting down details under each item in his plan, the technician may be guided by his own needs—the plan is for him to follow. Its effectiveness will ultimately be determined by the usefulness of the technical report to which it leads.

OUTLINING

The reasons for preliminary outlining are many. For the new writer, acquisition of the outlining habit is especially important. Each report will continue to be a troublesome chore unless the writer early develops work habits which will permit him to control the materials rather than be controlled by them.

Outlining begins to bring order out of chaos and forces the technician to see his way through the whole report before he actually begins writing. He is obliged beforehand to determine the relationships between the many bits and pieces of information that are swamping him. The outline and preliminary plan provide a check point for contact and coverage of everything necessary. Many details which might otherwise be forgotten cannot easily be overlooked. The writer is permitted to concentrate on selection and organization of materials without actually being concerned about matters of composition, which come later.

There is no formula to follow in the earliest stages of outlining, but broad general categories of information certainly begin to suggest themselves logically to any worker. The early or preliminary part will surely want to introduce the reader to the subject. What are the details of background at the writer's disposal that will quickly make the reader familiar with the subject at hand? As ideas start to fall into place, what must make up the central portion of the report? What materials will be used? What methods will be employed?

Following this, what results are anticipated? What conclusion can be drawn and what recommendations offered? During preliminary outlining the writer is actually selecting the list of main headings which his report will likely follow. He is sifting the smaller pieces of information—his raw materials—and placing them under appropriate headings and into logical sections. He is grouping topics under headings and subheadings so that he himself can begin to see how they all hang together. He is relating subgroups of information to one another and reclassifying them so that all items are accounted for in some kind of total relationship. Throughout this entire preliminary process the writer is keeping his eye on the objective of the report which he has clearly set for himself in terms of a particular audience. He is always thinking of each large or small detail with respect to where it will best fit—under what heading and section—in order to fully accomplish the objective at hand. There are always the major groupings and subgroupings to be worked and rearranged as he proceeds with this preliminary outlining.

As with many aspects of technical writing, the individual will develop his personal techniques for preliminary outlining. At the early stage it is perhaps best not to be bound to rigid formulas and procedures. Nonetheless, there are traditional formats for outlining which have proved greatly useful to generations of writers. The new writer should be familiar with the most widely accepted outlining mode. If he chooses to reject it, he will know what he is breaking away from and be able to weigh the advantages of his own against the traditional way.

Classic outline form is as follows:

I. Main headings
   A.⎫
   B.⎭ Main subheadings
      1.⎫
      2.⎭ Subsidiary headings (first degree)
         a.⎫
         b.⎭ Subsidiary headings (second degree)
            (1)⎫
            (2)⎭ Subsidiary headings (third degree)
               (a)⎫
               (b)⎭ Subsidiary headings (fourth degree)

Immediately clear is the fact that this outlining form permits the writer to lay out his materials in a total relationship of logical connections. The main ideas are shown not only in relation to other main ideas but in relation to the appropriate main subheadings and subsidiary headings of which they are themselves comprised. Elements misplaced, out of order, overlapping, overlooked, or illogically sequenced can be spotted. The larger as well as the smaller units are broken down into analytical parts and placed in balance and symmetry against one another.

Various forms that represent a departure from the classic, traditional outline have developed. The simplest and perhaps crudest system depends almost wholly upon underlining, indenting, spacing, numbering, and capitalizing headings and elements of the report for emphasis and variation. At their best, reports so "outlined" fail to show the full, logical relationships among the parts, and, at their worst, are repetitious and tend to bombard the reader with so many typographical gimmicks that as he becomes exhausted by them, each succeeding one fails to stand out from the preceding.

A more acceptable substitution is the so-called numerical system, which simply permits the material to be presented in the following manner:

    1.
        1.1.
        1.2.
            1.2.1.
            1.2.2.
        1.3.
    2.

While reports have been effectively presented in this mode of outlining, the system is often self-defeating because it breaks elements down into too many tiny bits and pieces. Relationships among the fragmented, multiple parts tend to become obscured as the reader loses sight of the overall picture. The writer tends to break every sentence down into a numerical point or subpoint, whether it is logically necessary for him to do so or not. In other words, both the writer and reader tend to lose sight of the organic relationships among the parts as the seemingly limitless system of numbering continues. What appeared to be an easy way out often ends up in complication and confusion. One tends to come back to the classic outline form simply because the various substitutes for it appear not to be as effective as the original.

a supplementary view...

HOW TO START TO WRITE A REPORT *By J. Packard Laird*

From Product Engineering, *January 23, 1961. Reprinted with permission from* Product Engineering, *Copyright McGraw–Hill, Inc., 1961.*

Some people read to find things out; some people write because they have something to say. Why shouldn't these people get together?

If you can read these words, you have already been taught in school all about how to write, including writing reports. The fact remains, most reports make dull reading and are a nuisance to write.

ANSWER THE FOLLOWING QUESTIONS:
(If your answers do not satisfy you, . . . read on.)

Have you ever enjoyed writing a report?

Has anyone ever read a report of yours with interest and excitement?

When you are writing a report and get stuck or bored, do you know how to get vigorously started?

Would you like to be a successful report writer?

Assume You Have What You Need

Imagine that the completed report is in front of you. You now can imagine yourself examining it in detail, reviewing its structure, content, and style. Since it is a good report, all you have to do is put down on paper what you reviewed in imagination. This sounds too simple, and you probably feel that you couldn't make it work anyway. Don't judge too soon. Don't forget, you are not learning how to write, you are learning how to enjoy being a successful report writer.

You now can shift your attention and try to answer such questions as: who will read the report; who will take action as a result of reading the report; will it be worth your time to prepare the report; will it really make any difference? To answer these questions you need to know how your report will affect others. Will it change their thinking in some useful way?

Apply the rule again. Imagine that you have just been told that a man who read the report is starting to take action as a result of what he read. Now, in imagination, you can consider who the man is and what

THE PRELIMINARY PLAN 33

he is going to do. You can review in imagination all the ways in which your report influences this reader.

Each time you apply the basic rule, "Assume you have what you need," you are merely visualizing how the parts of your work fit together to make a well-organized whole. If you ever get stuck while writing a report, it is probably because you are not applying the rule. More likely, you probably are refusing to apply the rule. It sounds silly; it is almost embarrassing.

You can't apply the basic rule unless you are sure that you will be successful. But now we are talking about an attitude or a belief. A man's beliefs are the hidden forces which guide his life. All successful men know they can succeed in any reasonable undertaking; they assume that success will prevail. Why don't you assume that your report will be an effective and persuasive literary masterpiece? Why not assume that you will succeed?

Assume That Success Will Prevail

A wise person assumes that success will prevail, assumes that he has what he needs to make the success possible, and then acts appropriately to achieve the assumed success. To such a person all aspects of any problem become clear and appropriately interrelated.

It is all well and good to say, "Assume that success will prevail," but who can actually believe it? By what means does a person bring about success? Actually there are many people who secretly and unknowingly plan on failure, since in this way they can claim a near-miss of godlike success.

If you are going to apply this basic rule, apply it to first things first. You need success, therefore, assume that you have succeeded. (If you assume you will write an ordinary report, it will probably be one of the worse.) Assume your report will be an all-out success. Then apply the rule again and assume you know why it is so successful. In this way you will uncover all the features which contribute to this success.

You will find that this attitude that success will prevail comes easily when you learn to do two things: discriminate and relate.

A successful report contains information about things which differ yet are related. The report itself has sections which are different yet related. In order for a report to be successful, it must be related, both within itself and with respect to the business organization throughout which it is distributed.

You must have the attitude that success will prevail in order to discriminate and relate effectively.

Discriminate—Relate

When a man reads your writing, he will get information and mental impressions about you and the subject of your writings. Your job is to convey the precise message which you intended.

The secret lies in knowing how to do two things:

YOU MUST DISCRIMINATE.

You must find out how to distinguish the one idea which concerns you at a specific instant from all other ideas. There are always distinguishing features which will discriminate any one idea. Your writing at that one point will become specific and very much to the point.

YOU MUST RELATE.

In this way you reassemble the ideas, which were discriminated for clear identification, into a meaningful whole entity.

Consider a few illustrations of improper discrimination and relationship. You have a report on a new machine. You are reading a description which tells about the physical layout of the assemblies, when suddenly you discover that you are reading about how the machine functions or why it saves money. In another case you may be reading about a basic principle in mechanical engineering only to discover that at some past point the author switched without warning to a specific example. The confusion in these examples is typical of the failure to discriminate and relate. This is the cause of most bad report writing.

You may have realized that you cannot discriminate without relating. For example, the difference between "pencil" and "pen" cannot be established without relating each to "paper" by means of "writing." In this example you would not bring in nonessential similarities, such as "carried in pocket," because this would not aid in discrimination. You would, however, bring in differences such as "solid lead" *vs* "liquid ink" and "erasable" *vs* "cannot erase what has been written."

Your job is to relate and discriminate at the same time. You choose that relationship which helps to discriminate.

Note that everything you write or say either discriminates or relates. You may be describing an additional difference or distinguishing feature of some plan of action or you may be relating this plan to the probable success of the company. Try to distinguish between discriminating and relating, then combine the two so that they reinforce each other and lock together.

There are a number of distinct areas in which you must discriminate and relate. Each of these areas has its own peculiar requirements.

Proper discrimination and relationship are important within each area. It is even more important for you, as writer, to know which area you are dealing with and not jump from one area to another without orienting the reader quite carefully. These broad areas are listed below.

POINT OF VIEW AND MOTIVATION

You have reasons for writing; the reader has reasons for reading. You want to derive the satisfaction of producing a report which is well received. Many readers will give you this satisfaction if you will discriminate your own selfish motivations, eliminating as many as possible, and relate everything in the report on a reader-interest basis.

ORGANIZATION OF REPORT

Be able to discriminate:
  abstracts, résumé, summary, table of contents;
  appendix, exhibit, attachment;
  table, graph, figure, illustration, diagram, sketch, schematic;
  discussion, section.

SUBJECT MATTER AND ITS RELATIONSHIP

Be able to discriminate:
  ambition, belief, objective; theory, hypothesis, assumption, principle (for example, scientific principle *vs* moral principle);
  attitude, behavior;
  method, procedure, means;
  intend, outcome, results, significance of results;
  hope, plan, accidental discovery;
  data, information, facts, knowledge, understanding, wisdom;
  definition, description, example, illustration;
  opinion, conclusion, recommendation, proposal.

REQUESTS AND PROPOSALS

If you haven't got a proposition, you don't have anything to say. Your proposal should be specific. For example, a progress report should clearly indicate progress from a stated starting point toward a stated objective and should propose continuation of the work as planned or modification of plans for stated reasons. Informational reports should specifically state that they will be useful as reference material and should tell why. Some reports are spoiled by subtle demands for praise or recognition of the author at the expense of the reader. A successful report makes proposals which will be of benefit to the intended reader.

Writing is an art; there is nothing routine or mechanical about it. You will want to think about this matter of discriminating and relating, and then practice your best when you write. It takes real creative effort to combine theory, ideas, facts, and calculations into a good report. Why not try to make it a work of art while you are at it?

Attack Each Job as a Creative and Artistic Art.

Every artist works solely for his audience. A good artist learns how to use his tools. He discriminates between tools, and he relates the tools to the job at hand. He discriminates between audiences, and he relates his work of art to his audience. He feels a deep conviction that his chosen audience will appreciate his work of art.

The quality of an artist is measured in terms of the audience which is influenced by his work and how that audience responds.

The next time you write, consider it to be a creative and artistic act. Assume that success will prevail. If any part of the over-all presentation is missing or is hazy, assume you have what you need. You will then be able to create this difficult portion by systematically relating specific thoughts which are distinct yet connected. Because you have kept the reader in your mind at all times, you will tell your story to him in such a way that he stays interested. You will have condensed and clarified your writing to make all parts stand out in proper relationship. You will have created a work of art, pleasing to the eye, the mind, and the heart of the reader.

Chapter 4
METHODS OF GATHERING INFORMATION

The material for a technical report will be brought together in a variety of different ways. While the technical writer may find himself using one or another method more frequently than others, it is necessary for him to gain familiarity and competence with all techniques for gathering information. No report can be better than the information upon which it is based.

THE LIBRARY AND SEARCHING THE LITERATURE

Before you can write a technical report on any subject, you must know what has already been written. Technology advances through the accumulation of printed information, and because the knowledge sometimes grows very rapidly, you must be sure you are abreast of the latest developments before you start to write. You begin the process of gathering information by making a search of technical and scientific literature.

The more rapidly the technical writer learns to use the resources of the library, the more quickly will his reports gain authoritative value for the reader. While the technician will in everyday practice probably come to find himself using periodical literature most frequently, he must nonetheless develop a broad familiarity with reference materials of all sorts, and with the library holdings listed in the card catalog.

Three principal kinds of holdings may be found in a library: 1) a general collection of books; 2) a collection of reference works; and 3) a collection of periodicals, newspapers, magazines, journals, bulletins, and pamphlets.

The card catalog. The card catalog is an index to the entire library. Every book is listed, including reference volumes usually found on the open shelves of the reference room, and every bound magazine and journal.

The card catalog consists of 3 x 5 inch cards arranged alphabetically in drawers. There will usually be three separate cards in the catalog for each book: 1) an author card; 2) a title card; and 3) a subject card.

If the technical writer knows the surname of an author of a book or books on the subject he is going to write about, he can find full information by consulting the author card. Or if the searcher knows only the title of a book but not the author, he can seek out the title card, which will be placed alphabetically in the catalog. But at the beginning of a search of the literature, the technologist will probably know few if any specific authors or books on the subject he is preparing. He will therefore go to the subject about which he is seeking information where, with the aid of cross reference cards, he will be able to search out all of the available materials the library holds.

There follows a specimen author card from the card catalog:

Specimen Author Card, Card Catalog

```
T           Moore, Arthur Dearth, 1895–
19              Invention, discovery, and creativity [by] A. D. Moore.
M76         [1st ed.] Garden City, N. Y., Doubleday, 1969.

                xiv, 178 p. illus., ports. 19 cm. (Science study series)
            4.95

                Bibliography: p. [170] –172.
                Story House Prebound


                1. Inventions  2. Creativity  (Literary, artistic, etc.)
                I. Title. (Series)

            T19.M76                    600                  69-10970
                                                              MARC
                Library of Congress          [5]
```

METHODS OF GATHERING INFORMATION 39

The key to the information on the specimen card catalog author card above is:
a. T
   19 } Classification call number of the book.
   M76
b. Moore, Arthur Dearth, 1895– Name of the author, date of his birth; he is living.
c. Invention, discovery, and creativity by A. D. Moore. 1st ed. Garden City, N.Y., Doubleday, 1969.–Full title of book with author's name repeated, edition, place of publication, publisher, and date of publication.
d. xiv, 178 p. illus., ports. 19 cm. (Science study series) 4.95 Bibliography: p. 170–172.
   Story House Prebound–Description of book: fourteen-page introduction, 178 pages of text, illustrations, and portraits; it is 19 centimeters high (there are 2.54 centimeters in an inch); it is part of a series, selling at $4.95. There is a bibliography on pages 170–172, and the book is available in a special binding for use by libraries.
e. 1. Inventions. 2. Creativity (Literary, artistic, etc.) I. Title (Series)–The book is also listed in the card catalog under these subject headings, and there is also a separate title card for it as well as for the series of which it is part. (Arabic numbers are used to indicate subject headings and Roman numerals to indicate title headings.)
f. T19.M76–Library of Congress call number.
g. 600–Dewey decimal system call number.
h. 69–10970–Order number used in ordering the book.
   MARC
i. Library of Congress–Book has been housed in and cataloged by the Library of Congress.
j. 5–A printer's key to the card.

The title card is a copy of the author card, with the title typed in just above the author's name. It is filed in the catalog according to the first important word in the title.
Each of the subject cards is also a copy of the author card shown, with the subject typed in (usually in red) above the

author's name. Since, as we said, neither authors nor titles are known at the outset of searching the literature in preparation for a report, the subject listing is often the classification most helpful to the student.

Classification of books. In a large library, which may contain a million or more volumes, some foolproof system for locating each book is necessary. For this reason a library assigns each book a specific code symbol or call number; this symbol, consisting of numbers and letters, is found on the spine of a book and on all catalog cards referring to the book.

The first line of a call number is the classification number, which will be uniform in all libraries using the same system. The lower line or lines are the author number, a combination of letters and numbers that identify the precise location of a book on a certain shelf.

The two classification systems in common use are the Dewey Decimal and the Library of Congress. The technician should know the general nature of the system used in his library, and he should know in detail the categories that apply to his major field of study.

Dewey Decimal System

The Dewey Decimal system, as the name suggests, is based on numerals divisible by ten. The fields of knowledge are divided into ten general categories:

000-099 General works
100-199 Philosophy
200-299 Religion
300-399 Social Sciences
400-499 Philology
500-599 Pure Science
600-699 Applied Arts and Sciences
700-799 Fine Arts, Recreation
800-899 Literature
900-999 History, Geography, Travel, Biography

Each general category is subdivided by tens. For example:

500-509 Pure science
510-519 Mathematics
520-529 Astronomy
530-539 Physics
540-549 Chemistry

550-559 Geology
560-569 Paleontology
570-579 Biology
580-589 Botany
590-599 Zoology

Each of these fields is also subdivided. For example:

510 Mathematics
511 Arithmetic
512 Algebra
513 Geometry
514 Trigonometry

515 Descriptive geometry
516 Analytic geometry
517 Calculus
518 unassigned
519 Probabilities

Each specific category is subdivided further by adding numbers after the decimal. For example:

511   Arithmetic
511.1 Systems
511.2 Numeration
511.3 Prime numbers
511.4 Fractions

511.5 Analysis
511.6 Proportion
511.7 Involution, evolution
511.8 Mercantile rules
511.9 Problems and tables

Library of Congress System

The Library of Congress system, which is based on letters, is preferred by many large libraries because a book can be classified with a shorter symbol. The major categories are as follows:

- A. General works, Polygraphy
- B. Philosophy, Religion
- C. History, Auxiliary sciences
- D. History, Topography (except America)
- E. America (general), United States (general)
- F. United States (local), America (except the United States)
- G. Geography, Anthropology
- H. Social Sciences (general), Statistics, Economics, Sociology

J. Political Science
K. Law
L. Education
M. Music
N. Fine Arts
P. Language and Literature
Q. Science
R. Medicine
S. Agriculture
T. Technology
U. Military Science
V. Naval Science
Z. Bibliography, Library Science

Further subdivision can be illustrated with "S," one of the simplest categories:

S   General agriculture, soils, fertilizers, implements
SB General plant culture, horticulture, parks, pests
SD Forestry
SF Animal culture, veterinary medicine
SH Fish culture, fisheries
SK Hunting, game protection

Periodical literature. Current and contemporary technological and scientific information is reported and recorded in journals, magazines, and periodic papers. The most frequently consulted of these (depending on the nature of the library) are often freely available in current issue on open shelves in the periodical section. Older issues of periodicals are bound in volumes and shelved in the stacks or may be available on microfilm. The majority of libraries today in almost any geographic locale have excellent systems of interlibrary lending and exchanging so that a copy of almost any article not on hand may be readily obtained.

While all periodical publications held by a library may be located by title in the card catalog, most libraries have a separate catalog of periodical titles in order to expedite the search. The card will indicate the range of issues of the periodical held by that library.

METHODS OF GATHERING INFORMATION

In practice, however, the student will usually begin his search of periodical literature by consulting one of the general or special indexes. Scientific and technological articles are listed in these according to subject and, in some, by author and title as well. Listings are kept remarkably current by supplements six or eight weeks following publication. The student will therefore wish to consult not only the bound annual indexes for any period but the supplements as well.

The following listings of the general and some special indexes to periodical literature are given in abbreviated form. Since the lists are arbitrary here, they in no way aim to cover the entire range. For example, the special indexes touch only upon applied science and exclude almost all the guides in science, of which there are many. The intention is to suggest the possibilities to the student so that he may be stimulated to search out those special periodic and reference materials which apply specifically to his field—those which will prove invaluable to him while searching the literature and preparing the report.

Indexes to Periodicals

General

Poole's Index
Readers' Guide
Book Review Digest
International Index
New York Times Index

Special

Aeronautical Engineering Index
Aeronautical Engineering Review
Agricultural Index
Applied Mechanics Reviews
Bibliographic Index
Biography Index
Education Index
Engineering Index
Electronic Engineering Master Index
Index Medicus; Quarterly Cumulative Index Medicus
Index to Legal Periodicals

Industrial Arts Index; Applied Science and Technology Index; Business Periodicals Index
Public Affairs Information Service
Technical Book Review Index
Technical Data Digest
See also the various abstracts, such as *Biological Abstracts* and *Chemical Abstracts.*

Most-used science and technology periodicals. The number of scientific and technical periodicals published regularly in English and foreign languages is great. At best, any competent technologist or scientist may hope to become familiar only with those journals relating to his special interests. As a matter of some value, however, the New York Public Library, Division of Science and Technology, Reference Department, has made a one-year tabulation of its one hundred most-used periodicals. The study applies only to one library at one time and one place, and therefore may have little general relevance and applicability. Since it does, nonetheless, offer a selected listing of one hundred periodicals on some rational basis, the list is reproduced in the expectation that each technologist will recognize some titles he is himself familiar with. Libraries in teaching, business, or industrial settings— according to their special needs—will probably have on hand or available on loan some of the following listed periodicals.

NOTE: Letters in parentheses indicate weekly (w), biweekly (bw), monthly (m), bimonthly (bm), and quarterly (q).

1. Chemical & Engineering News (w)
2. Industrial & Engineering Chemistry (m)
3. American Chemical Society, J. (bw)
4. Engineering News-Record (w)
5. Oil & Gas Journal (w)
6. Aviation Week (w)
7. Scientific American (m)
8. Physical Review (bw)
9. Nature (w)
10. Analytical Chemistry (m)
11. Chemistry & Industry (w)
12. Chemical Society, London, J. (m)
13. Electronics (w)
14. Science (w)
15. Iron Age (w)
16. I R E Proceedings (m)

17. Journal of Chemical Physics (m)
18. Philosophical Magazine (m)
19. Modern Plastics (m)
20. Journal of Applied Physics (m)
21. Astronomical Soc. Pacific, Pub. (bm)
22. Steel (w)
23. Chemical Engineering (bw)
24. Review of Scientific Instruments (m)
25. Engineering (w)
26. Product Engineering (w)
27. Automotive News (w)
28. Petroleum Refiner (m)
29. Faraday Society, Transactions (m)
30. Journal of Chemical Education (m)
31. Chemical Week (w)
32. Journal of Physical Chemistry (m)
33. Petroleum Week (w)
34. Physical Society, London, Proc. (m)
35. Popular Science (m)
36. Chemische Berichte (m)
37. Electrical Engineering (m)
38. Optical Soc. America, Journal (m)
39. Amer. Inst. Elec. Engrs. Trans. (q)
40. Journal of Polymer Science (m)
41. American Soc. Mech. Engrs. Trans. (q)
42. Oil, Paint & Drug Reporter (w)
43. Acoustical Soc. America, J. (m)
44. Engineer (w)
45. American Dyestuff Reporter (bw)
46. Machine Design (bw)
47. Franklin Institute, Journal (m)
48. Z. fur Anorg. u. Allgem. Chemie (m)
49. Rec. de Trav. Chim. du Pays-Bas (m)
50. American Ceramic Soc., J. (m)
51. Liebig's Annalen der Chemie (m)
52. American Mathematical Monthly (m)
53. Paper Trade Journal (w)
54. Mechanical Engineering (m)
55. J. of Scientific Instruments (m)
56. Journal of Biological Chemistry (m)
57. Motor (England) (w)
58. American Journal of Science (m)
59. Electrical World (w)
60. Chemical Reviews (bm)
61. Petroleum Engineer (m)
62. Angewandte Chemie (m)
63. Zeitschrift fur Physik (m)
64. Factory (m)
65. Nucleonics (m)
66. Electronical Manufacturing (m) (now Electro-technology)
67. Journal of Metals (m)
68. Textile World (m)
69. Rubber Age (New York) (m)
70. Zeitschrift fur Naturforschung (m)
71. Journal of Organic Chemistry (m)

72. Zeitschrift fur Physik. Chemie (m)
73. American Machinist (bw)
74. Reviews of Modern Physics (m)
75. Scientific Monthly (m)
76. Engineering & Mining Journal (m)
77. Zeitschrift fur Elektrochemie (m)
78. Physica (m)
79. Automotive Industries (bw)
80. Isis (m)
81. QST (m)
82. Biochemical Journal (m)
83. United States Naval Inst. Proc. (m)
84. TAPPI (m)
85. Bull. of the Atomic Scientists (m)
86. Analyst (m)
87. Astrophysical Journal (m)
88. Naturwissenschaften (bw)
89. Hardware Age (bw)
90. Electrical Merchandising Week (w)
91. Power (m)
92. Corrosion (m)
93. American Journal of Physics (m)
94. Sewage Works Journal (m) (now Water Pollution Control Federation, J.)
95. Annals of Mathematics (m)
96. Autocar (w)
97. American Mineralogist (m)
98. Chemical Engineering Progress (m)
99. Air Cond., Htg. & Refrig. News (w)
100. American Water Works Assn. J. (m)

Reference works. There are general and special reference works of great value to all technical writers. These important sources of information will be consulted throughout all stages of writing any report. But they appear especially helpful at the beginning of the information-gathering process and in the preparation of the preliminary bibliography.

The resources of different libraries vary. Writers seeking special information may have to depend upon several library collections. The technical writer will gradually become familiar with those reference works that are most important to his particular field and will determine which libraries in his locale he must use for special purposes. The same will hold true for periodical guides and technical journals.

Since general and special reference works are numerous, it is unlikely that many writers will become familiar with them all. Any selected list of them will be incomplete. But the following extremely abbreviated listing may serve merely to suggest the ranges and categories of standard reference tools and to provide some specific titles (in shortened form) of usefulness to the beginning technical writer.

Guides to Reference Books

Cook, M. G. *The New Library Key.* New York: H. W. Wilson, 1956.
Murphey, Robert W. *How and Where to Look It Up; A Guide to Standard Sources of Information.* Consultant: Mabel S. Johnson. Foreword by Louis Shores. New York: McGraw-Hill Book Co., 1958.
Russell, H. G., R. H. Shove, and B. E. Moen. *The Use of Books and Libraries.* Minneapolis: University of Minnesota Press, 1958.
Shores, Louis. *Basic Reference Books.* Chicago: American Library Association, 1954.
Winchell, C. M. *Guide to Reference Books.* Chicago: American Library Association, 1951. Supplements, 1951–52, 1953–55, 1956–58.

General Encyclopedias

Columbia Encyclopedia. 2nd ed. Ed. by William Bridgwater and Elizabeth J. Sherwood. New York: Columbia University Press, 1950.
Encyclopedia Americana. New York: American Corporation, 1956. 30 vols.
Encyclopaedia Britannica, 14th ed. New York: Encyclopaedia Britannica, Inc., 1956. 24 vols.
New International Encyclopedia. 2nd ed. New York: Dodd, Mead and Company, 1914–16. 24 vols. Plate revision, 1922. Supplements, 1925, 1930.

Dictionaries, Word Books

Dictionary of American English on Historical Principles. Ed. by Sir W. A. Craigie and J. R. Hulbert. Chicago: The University of Chicago Press, 1936–44. 4 vols.
Fowler, Henry W. *Dictionary of Modern English Usage.* New York: Oxford University Press, 1934.
New Standard Dictionary. New York: Funk and Wagnalls Company, 1952.
Oxford English Dictionary. Ed. by A. H. Murray *et al.* Oxford: The Clarendon Press, 1888–1933. 10 vols. and supplement. Reissue, corrected, 1933. 12 vols. and supplement. The original issue is known as *New English Dictionary.*
Perrin, Porter G. *Writer's Guide and Index to English.* 3rd ed. Chicago: Scott, Foresman and Company, 1959.
Roget's International Thesaurus. New ed. New York: Thomas Y. Crowell, 1946.
Webster's Dictionary of Synonyms. Springfield, Massachusetts: G. & C. Merriam Company, 1942.

Webster's New International Dictionary. Unabridged. Springfield, Massachusetts: G. & C. Merriam Company, 1954.

Atlases

Columbia-Lippincott Gazetteer of the World. New York: Columbia University Press, 1952.

Commercial Atlas. Chicago: Rand, McNally and Company. Issued annually.

Cosmopolitan World Atlas. 2nd ed. Chicago: Rand, McNally and Company, 1951.

Encyclopaedia Britannica World Atlas. New York: Encyclopaedia Britannica, Inc. Frequently revised.

Webster's Geographical Dictionary. Rev. ed. Springfield, Mass.: G. & C. Merriam Company, 1955.

Technical Dictionaries, Handbooks, and Encylopedias

American Civil Engineers' Handbook
American Electricians' Handbook
American Machinists' Handbook and Dictionary of Shop Terms
Civil Engineering Handbook
American Standard Definitions of Electrical Terms
Baughman's Aviation Dictionary and Reference Guide
Casey Jones Cyclopedia of Aviation Terms
Chambers' Technical Dictionary
Chemical Engineers' Handbook
Chemical Engineers' Manual
Dictionary of Technical Terms
Engineering Encyclopedia
General Engineering Handbook
Glossary of the Mining and Mineral Industry
Handbook of Non-Ferrous Metallurgy
Handbook of Plastics
Heating, Ventilating, Air Conditioning Guide
Heating and Ventilating Engineering Databook
Iron and Steel Pocket Encyclopedia
McGraw-Hill Encyclopedia of Science and Technology
Mechanical Engineers' Handbook
Metallurgists' and Chemists' Handbook
Metals and Alloys Dictionary
Petroleum Almanac
Petroleum Data Book
Petroleum Dictionary
Petroleum Facts and Figures
Plastics Dictionary

METHODS OF GATHERING INFORMATION 49

Radio Engineering Handbook
Radio Engineers' Handbook
Refrigerating Data Book and Catalog
Society of Plastics Industry Handbook
Technical Dictionary of Terms Used in Electrical Engineering
Tool Engineers' Handbook
Welding Handbook

Yearbooks and Directories

American Business Directories
Americana Annual
Annual Register
Britannica Book of the Year
Economic Almanac
Engineering Societies Yearbook
Handbook of Scientific and Technical Societies and Institutions of U.S. and Canada
Information Please Almanac
International Industry Yearbook
MacRae's Blue Book
New International Yearbook
Statesman's Yearbook
Statistical Abstract of the United States
Sweet's Catalog Service
Thomas' Register of American Manufacturers
Whitaker's Almanack
World Almanac

During the preliminary stages of gathering information for a report, the writer may wish to gain a general view of the unfamiliar subject he will deal with in his report. An article in one of the special encyclopedias may provide such an overview. While developing his information and writing his report, the technician may need to define a term, establish charts, review a technical process unfamiliar to him, or quickly and easily locate many different kinds of facts and isolated pieces of information. It is to the encyclopedias, dictionaries, year books, guides, directories, and other invaluable reference works that he will go for these fragmentary but necessary items.

Chapter 5
PRELIMINARY BIBLIOGRAPHY

From the search of the literature comes the preliminary bibliography which each writer must prepare. This bibliography represents a listing of all library materials which, before direct work on the report at hand actually begins, appear to have a bearing on the subject. These are the books and articles the technical writer will survey, discarding some and using others, before he undertakes his own original work. With a preliminary bibliography to work from there is a solid foundation for the report.

Searching the literature and preparing the preliminary bibliography occur simultaneously. As the writer comes across materials in the card catalog or periodical indexes that appear to be potentially useful, he notes all the pertinent information on a filing card. It is best to complete the preliminary bibliography systematically before attempting to examine any of the material. Preparation of the research note cards, the next step, may then also proceed uninterruptedly.

The preliminary bibliography card prepared by the writer lists whatever information may be needed: 1) to obtain the book or article from the library stacks or open shelves; and 2) to incorporate the book or article into footnotes and a final bibliography in the event that the material comes to be used and cited in the report. To account for all this the card prepared by the technician will usually look like the following:

> Knepler, John.
> "Cross-modulation + IM in
> Receiver R.F. Amplifiers,"
> <u>Electronics World</u>
> March 1970, Vol. 83, No. 3.
>
> (13)
>
> TWTN
> (S&T.)   pps. 55-58

The basic data for the preliminary bibliography card can be obtained from the card catalog or, as in this case, one of the guides to periodical literature. The library call number noted in the lower left corner must be obtained either from the card catalog or the periodical catalog, found separately in most libraries. The number in the upper right corner is that assigned to this bibliographic entry by the individual report writer; it will prove useful during the note-taking process that follows. Presented thus, the preliminary bibliography card contains all the information for obtaining the article, citing the reference in a footnote, and listing it in the final bibliographical list accompanying the report.

Since the method of gathering raw library material for the technical report is such a major integral phase of preparing the final document, the technician may make his writing job more pleasant by developing orderly procedures. A good many technicians prefer to use ordinary 3 x 5 inch file cards for the preliminary bibliography and to arrange these alphabetically by author's surname. A separate card is prepared for each book from the general collection, each item from the periodical literature, and each reference source that appears promising. So that it will not be necessary to return either to the catalog or the indexes, all relevant information is noted. Depending on the work habits and preferences of the individual, it is probably best to complete the preliminary bibliographic search before undertaking preparation of notes. Additional sources, of course, will be encountered and added to the bibliography as the examination and reading of material gets under way.

RECORDING LIBRARY NOTES

By the stage in the preparation of the report at which note-taking begins, the technical writer has come a long way. He has determined the objective of his report and fixed his specific subject firmly in mind. He understands the aim of his assignment and has a clear knowledge of the nature of his audience in relation to the purpose of the report. He has prepared a tentative preliminary outline or plan for the report, understanding that such a plan is subject to alteration as his research proceeds. And he has searched the literature and gathered a preliminary bibliography. Eventually he will write his report while using his outline, bibliography, and note cards, and from that draft he will prepare his final manuscript for presentation.

The value of the substance of the report will depend greatly on the quality of the library source material used and the ability of the writer to record it effectively. As the student proceeds systematically to obtain and examine each item in his preliminary bibliography, he will first evaluate the quality of the material before attempting to record it. Is the journal well known? Do the author and his affiliation appear reliable? Are the journal and/or author generally recognized as authoritative? Is the edition of the book the latest one available? Does this source appear to be the most dependable one for the information sought? While only time and experience in a particular technical field can make it possible to answer these questions readily, the student may at the outset seek help from his peers, librarians, professors, and supervisors. For technical books one may seek critical reviews by consulting the *Technical Book Review Index*.

It is a poor habit to use library material uncritically simply because it is at hand and available.

Books and articles, once critically evaluated for their reliability, seldom need to be read entirely. Contents, prefaces, and indexes of books help the student locate the chapters or sections he requires. Abstracts and section headings are similarly useful in articles. Each researcher gradually develops his own methods for efficiently working printed materials. Ability to read well is intimately allied with ability to take notes.

Library notes may be recorded in many ways, but the conventional method employs file cards or paper sheets of uniform size. Each card contains a single note with a heading keyed to a part of the preliminary outline. Cards are thus arranged according to the place they will take in the organization and development of the report. An acknowledged time-saving device is to key each note card to the appropriate bibliography card. In that way bibliographic information need not be duplicated many times but is easily available to the writer by cross reference. Such information will clearly be needed in a paper using footnotes and bibliography, the standard practice. Specific pages referred to usually appear on the note card.

A note represents a digest or abstract in the writer's own words of the essential information drawn from the source. Skill must gradually be acquired in recognizing the basic and relevant data. Such skill grows with the discipline of writing, since the writer becomes increasingly conscious of organizing and developing his own material and thus recognizes how other writers present their work by plan. Writers of integrity are constantly aware of the need to credit their sources fully. Any failure, of course, to acknowledge source material is a serious offense called plagiarism. Where the borrowed idea is expressed in the writer's own phraseology, citation of the source is sufficient. Where, for whatever reason, the idea is presented in the phraseology of the source, quotation marks are required. Since the implications of note-taking are so varied and important, the basic technical writer should strive early to distinguish clearly and separately on each card among what is his own summary and paraphrasing, what is a direct quotation, and what is his own evaluative commentary and critical interpretation of the recorded material. If the note card does not readily yield up these separate categories of information during the final writing stage of the report, considerable rechecking and rewriting may be called for. If intelligence is the ability to learn from experience, every technical writer will allow this frustrating need to recheck to happen no more than once.

A note card will probably look something like the following:

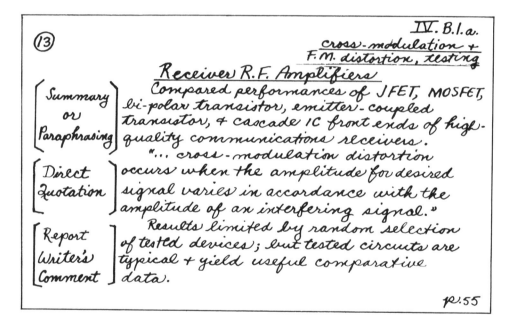

In the upper left corner is the report writer's key to his own set of bibliography cards. In the upper right corner is the key to that section of the writer's working outline into which the recorded information will ultimately be incorporated. Since preliminary or working outlines may change, the writer has also keyed the note to a subject. Expansion or reorganization of the plan is sometimes easier with such subject headings on the note cards. The page-number at bottom right is the page from which the direct quotation was taken.

OTHER METHODS OF GATHERING INFORMATION

While most of the information a technician gathers for reporting may come from printed library holdings, some reports will require performance of laboratory or field experiments for basic information; others may need preparation of a questionnaire or the conducting of an interview. No single method of information-gathering will suffice indefinitely for all the reports one may undertake.

Explanation of each of these methods is outside the subject of this book. Technical people, however, are usually familiar with laboratory or field experimental procedures or know where to obtain guidance. Those to whom scientific process is not strange know that an experiment consists of an arranged, systematic series of objective observations under controlled conditions, designed to answer particular questions. These procedures must be repeatable and verifiable to be considered scientifically valid. Safeguards must be taken against variables and bias, and controls must be installed against these.

Preparation of questionnaires or the conducting of interviews in person are useful modes of information-gathering in selected instances. Considerable skill is needed in determining the kind of question—free-answer, two-way, or multiple choice—and in framing the order of questions. Special practice and study of the technique are needed if the information developed is to be fully useful to the report writer. Appropriate and representative respondents must be chosen, and knowledge of how to interpret results and draw conclusions from them is needed, while allowing for a margin of error. No writer without expertise in questionnaire preparation—gained through extended study and experience—should even undertake the process. The same restriction applies to conducting a series of personal interviews for information. Only the most seasoned of technical writers, practiced in special skills beyond both technology and writing, could be expected to prepare a report based on interview-information. These additional methods of gathering information are mentioned here merely to acquaint the student with their existence, possibility, and challenge for the future.

Chapter 6
THE JOB TECHNICAL WRITING DOES

Technical writing in its many varieties may be said to be concerned with five different general activities: 1) explaining how to do something; 2) explaining how something works; 3) explaining what something is; 4) explaining how things are put together; and 5) explaining why certain causes produce certain effects.

The next two chapters indicate how various methods of organizing and developing material for explanation serve the writer in the fulfillment of his main purpose. In explaining how to do something—that is, giving instructions for carrying out a procedure such as installing new windshield wipers—or in explaining how something is done—that is, how a process such as manufacturing plastic buttons is carried out—the technician will both organize and develop his information in special ways. He will probably choose either a chronologic or spatial method of organization and depend chiefly upon a process pattern of development. Emphasis will be upon the sequence of actions one person performs—or the sequence of actions various persons perform in following the instructions or in carrying out the process.

In describing a mechanism and explaining how it works, however, the writer may adopt a somewhat different mode of organizing and developing his writing. An inductive pattern of organization would probably be used in explaining how a jig saw works, and both description and definition would be required in developing the explanation. The main general intention of a particular piece of technical writing—in doing one of the five jobs mentioned above—will determine the means of setting up the writing and carrying it out.

ORGANIZING THE EXPLANATION

Many technical writers say that finding a means of organizing material for a report is the most difficult problem. They have the necessary information for the report in hand, and they know what the report is supposed to do, but they feel stymied about a way to organize it so that it will make sense for the reader. Once writers become aware of the various methods which permit them to arrange material according to logical patterns, the problem of organizing the report begins to appear much less difficult.

Chronologic order. Organizing material for a technical report according to a time-pattern is often the simplest and best way. The writer arranges things in the order in which they occur. He follows step-by-step order. The sequence of steps or happenings themselves determine the order in which they will be arranged for reporting. The natural order is allowed to prevail.

Whatever failures result in trying to make clear a process by chronologic order usually are traceable to lack of consideration for the nature of the particular audience. The writer forgets how much or how little the reader may know about the subject. When this happens the report neglects to make clear all of the specific details of the chronology or fails to explain properly all the links between steps.

There are several other possible difficulties in the otherwise relatively trouble-free chronologic method of arrangement that the technologist should be aware of. First of all the writer must remember that each step in a chronology is clear and separate but that no one step is more important than the total process.

This is to say that the writer and the reader should keep in mind the full process, of which any one step is but a single part. By referring both backward to preceding and forward to following steps, the writer can avoid losing sight of his true reason for writing, namely, to make the whole process clear.

While each step in any chronologic sequence is an essential link in the chain, some steps are more important to the total process than others. The technical writer should exercise judgment in treating in fuller or lesser detail any one step in accordance with what he considers to be the reader's needs. Chronologic arrangement does not imply that every step treated in order of time must be equally developed. If the writer knows that his particular audience is already fully versed in some step of the technology, it would be wasteful for him to develop this area of the report.

In each step of a chronology it is possible to emphasize a point of view corresponding to the interest of the audience. A certain substance, for example, might find technical application in both the textile and photographic industries. A writer preparing a report on the material for textile dye technicians would necessarily direct attention at each step to that particular technology. But he might wish to suggest that a property of the material producing a certain useful result in textile dyeing would produce a similar effect in photography technology, where quite different circumstances prevail.

A variation of the chronologic method of arranging material may also prove useful to the writer on occasion. This is reverse chronology. For reasons of emphasis the writer might choose to begin his process-explanation with the important final step and then to work back systematically to the first step in time. If the step of critical importance in the chronology occurs some place mid-way in the process, the writer can nonetheless start out with it and work forward or backward to the explanation of the other steps. With any variation of this sort, the writer must weigh advantages and disadvantages which may result for the reader.

Length, content, and complexity of instructions vary considerably. In following a chronologic step-by-step pattern of paragraph organization, the writer would necessarily have a good knowledge of his subject matter; try to satisfy the needs of his intended reader; be complete and accurate; keep to the point; and avoid gaps in the sequence of explanation.

Technical instructions most often employ the second-person pronoun *you* and the imperative or command voice of the verb.

For example: You first *place* the lamp across a workbench. Then, *remove* the base by holding the arm assembly. *Use* a plier-wrench.

Once the explanation is started, the pronoun *you* is established and understood and may be omitted, with only the imperative, action, command voice of the verb being used.

Instructional explanation often commences simply by identifying the operation to be explained:

Instructional explanation often commences simply by identifying the operation to be explained:

The purpose of this instruction sheet is to explain how . . . .

Following this, it is customary to state for whom the instructions are intended and when, where, and why the instructions are useful:

The homeowner needs to know how to install ____ when attaching the ____ so that the mechanism will function smoothly.

The first part of the explanation also conventionally includes a listing of the materials, tools, and equipment required. Definitions of any terms which may be unfamiliar are also offered here.

Thereafter follows the main part of the instructions in the step-by-step order discussed above. Sufficient details are provided at each phase, and if special caution to avoid error at any step seems appropriate, this is given. It is frequently useful—following the main, chronologic instructional presentation—to conclude the explanation with a summary or recapitulation of the chief steps. At the end, the writer may provide a general comment on the value of the operation, refer to alternate modes for carrying out the procedure, or conclude with some helpful suggestion.

Example: Paragraph Organization by Chronologic Pattern

ATTACHING MOWER HANDLE AND CONTROLS

19" H.P. MODEL

1. Attach lower handle to mower. It will snap over handle pivots by spring action. Secure in place by inserting hair pin cotters as shown.

2. Adjust handle height and angle by selecting either of 3 corresponding holes in handle latches and snap onto handle side pins.

3. Attach control panel to upper handle with the 4 self tapping screws provided. Run control cabling under lower handle as shown.

4. Attach upper and lower handles together with handle bolts and lock nuts.

5. Secure throttle cable to lower handle by snap-on clip.

6. Use three handle clips to retain wire harness.

7. Install battery POW-R-PACK on the control panel by inserting two (2) tabs into slots provided on the control panel; lift battery POW-R-PACK straight up and snap into place.

8. Insert connector on the wiring harness into the receptacle on the battery POW-R-PACK. The connector should be all the way in on receptacle.

9. See battery charge instructions on page 14.

21" S.P. AND H.P. MODEL

1. The upper and lower handles are assembled at the factory.

2. Follow the above steps 1 and 2 (19" H.P. Model).

3. Attach the control panel to the upper handle with the four self-tapping screws provided. Run control cabling over and on top of handle, as shown on page 4, to prevent kinking of the cables when the handle is folded for storage.

4. Secure cable or cables, depending upon model, to upper part of handle with handle clips provided. See page 4 for location of handle clips.

5. Route the 3 handle clips thru the slots provided in the wire harness and secure to handles as shown on page 4. NOTE: Wire harness should be secured to the topside of the handle.

6. Follow the above steps 7 thru 9 (19" H.P. Model) for attaching the battery POW-R-PACK.

Example: Paragraph Organization by Chronologic Pattern

Place the lamp across a workbench or table and disconnect all wire splices at the base so that it will be free to turn. Remove the base by holding the arm assembly with one hand and turning the base counter-clockwise with the other. This will separate it from a length of ¼-inch pipe which runs to the arm inside the pipe standard. After the base has been taken off, the standard and all other structural components will slide off the pipe. Use a plier-wrench or small pipe wrench to unscrew the locking ring joining the pipe to the arm assembly. Lay all metal components out on the bench (Fig. 16) for cleaning with an appropriate metal polish before reassembly.*

Spatial order. In ordering report material by space the technician moves from place to place or from part to part or from area to area. Spatial order permits reporting of the subject in terms of units which comprise it—the units are broken apart, laid out on the table, and reported by the idea, the responsibility, or the function of each as seen from a controlling vantage point.

Spatial arrangement by idea or area of responsibility will serve the technical writer only rarely. But basic technical writing frequently uses the function of each unit in space as the basis for arrangement. The chief challenge for the writer is to determine the most logical system for laying out the parts for analysis and explanation.

Parts move up or down or from side to side or in circular or other motions in space. Objects are stationary or they are

---

* From *The Popular Mechanics Home Book of Electrical Wiring and Repair* by H.P. Strand. Copyright 1960; 1964 by H.P. Strand.

moving. Direction of movement depends upon the position for viewing. This controlling point of view for establishing the relationship in space of the viewer and the object must be established by the writer as the basis for his reporting.

As with chronologic order, spatial arrangements which follow natural courses are usually best. A tiny object which must be viewed through a microscope to be seen is no less challenging to report or describe than a large die-cutting machine. But with the microscopic object the reader is by the nature of things obliged to take up the same position for viewing as the writer. Seen through the microscopic lenses the movement of the part from left to right or vice versa will necessarily be viewed similarly by writer and reader, if conditions are equal.

With larger objects having height and depth and width and consequent multiple planes, the writer must fix a position for viewing corresponding to the technical needs of the reader. If the report explains how to position parts for packing, the writer views the casing from above or from the left in the same manner that the packer must proceed. If the report explains proper oiling procedures for maintaining certain moveable parts, the writer assumes a position in relation to the machine that the oiler must assume—whether underneath or on top of or whatever—for carrying out the procedure.

Arrangement of report materials is logical if the writer establishes his position in space in relation to the object explained. Once the reader is made to understand where in spatial relation to the object the writer has placed himself, the report can proceed. If spatial relations shift—that is, if either writer or object shifts positions in the course of reporting—the reader must be appropriately informed.

The spatial pattern of paragraph organization works usually with description as the mode of paragraph development. Size, shape, weight, color, texture, material—these are the elements around which descriptive explanation is organized. From the vantage point of using the spatial pattern of paragraph organization in terms of parts of a mechanism, the writer is concerned mainly with construction of the device—the major components perhaps first, and then each of these in parts.

It would be misleading to suggest that explaining how something works from point of view of function, physical characteristics, or parts is a separate process. All three arrangements complement one another and are frequently presented in parallel fashion for the reader's full understanding. Yet the logical writer is fully aware always of what frame of reference he is using at any point and thus maintains control over and gives direction to the descriptive explanation as it develops. So long as elements are kept separate and used in logical order, no problem will occur. The writer must know beforehand whether a general description of a mechanism in overall view is wanted, or a specific description involving minute characteristics and details—code numbers, brand names, handbook terms for all parts, etc. Which is wanted will be largely determined by the purpose of the writing and the intended audience.

Example: Paragraph Organization by Spatial Pattern

> **NIKON'S BELLOWS** Focusing Attachment PB-4, with its front lens movements, is one of the most versatile macro systems around. The bellows' length is a generous 143-mm (about 5¾ in.); its shortest extension is 43-mm or a little less than 2 in. The front standard (with the lens) and the rear standard (holding the camera) can be moved back and forth independently. This serves several purposes.
>
> First, it enables the use of short retrofocus lenses, such as the 24-, 28-, and 35-mm Nikkors for high magnification work up to about 4.4X, either in normal or in reversed position with the BR-2 ring (the 24-mm can only be used the latter way). For additional magnification with some lenses, extension tubes may be added to the body.
>
> In ordinary bellows constructions, the rear member cannot be moved forward, and therefore the bellows track very often won't let the object get close enough to the lens. With this unit, it's no problem since you can move the lens standard all the way forward, and then the camera standard, to establish correct focus.
>
> The ability to shift both back and front standards can also come in handy when trying to establish the best weight balance in certain situations. If, for example, you have exhausted the movement offered by the lower sliding rail assembly, and the setup is a bit unbalanced, you can shift both lens and camera body to restore equilibrium.
>
> Many photographers aiming for a given magnification first establish

the correct lens-to-film distance. Then they set the camera up at approximately the correct film-to-object distance, and focus by racking the entire bellows assembly back- or forward, using the lower sliding rail. Here, too, independent back and front movements can make things easier for you.

Even though the movement of the lower sliding rail is quite generous, mistakes or peculiarities in setup may demand some impossible change in film-to-object distance. That's when the harried photomacrographer will appreciate the ability to move the body to the correct distance, then the lens, and finally fine focus with the sliding rail.

The front movements include a lateral or side-to-side shift of about 10-mm, and a side swing of about 30 degrees. The swinging feature can give extra depth-of-field in certain situations. The side shift is useful in helping to center the image, especially when using the accessory side copier; it allows you to copy a corner area of a slide.

The copier attaches to the front of the bellows. It accepts slides in all commercial 2x2 mounts, as well as strips of film. A Nikon Exposure Dial for slide copying may be used with various light sources and film types.

The bottom plate of the bellows' sliding rack has a 1/4x20 tripod socket, plus a series of holes for mounting it correctly on the stand of the Reprocopy Outfit Model PF.

The Nikon or Nikkormat body can be rotated to vertical, even when the rear standard is all the way forward on the track. The smooth-working, reasonably-sized knobs all have locks.

The instruction book has a comprehensive table of magnification ratios for the various Nikon lenses from 24- to 300-mm, along with other lens information rarely given by other manufacturers with their close-up equipment. It covers such data as the optimum aperture for close-ups, below which sharpness begins to deteriorate; which Nikon lenses are not suitable for copying and which really are; and under what conditions vignetting might take place.

Because of the number of lenses that Nikon recommends for bellows use, the usual magnification and exposure factor figures are not engraved on the track. Instead, a millimeter scale is furnished. Exact magnification ratios for any lens are easily figured by using the simple arithmetic recommended in the instruction manual. There is also a scale in the manual: when viewed through the finder, it is useful in figuring out image ratios.

The bellows has no provision for either a double-cable release or internal automation with any of the lenses.

Exposure compensation for extension is made automatically in Nikon and Nikkormat cameras with through-the-lens meters. All readings are taken the stop-down, match-needle way. Nikon automatic lenses operate manually.

Clear directions are given for Nikon and Nikkormat cameras without through-the-lens meters. The manual also covers differences in pupillary magnification of various Nikon lenses.

The instruction book was evidently printed before the availability of the 105-mm $f/4$ Nikkor-P, which is specifically designed for close-up shooting. This lens has a preset diaphragm with click-stops at 1/3-stop increments down to $f/32$. It focuses from infinity to a ratio of about 1.2:1, so it is suitable for a wide range of general photography as well as close-ups. . . .*

Example: Paragraph Organization by Spatial Pattern

Accordingly, power makes a round trip, or loop, on the switch cable, traveling from ceiling outlet to switch and back again. At the start, the switch loop's white wire moves the current from the black wire of the feed cable and delivers it to the switch. From there, the loop's black wire brings it back to the same outlet box, but this time it will be connected with the black wire lead of the fixture. As in all other wiring procedure, the white wire of the feed cable goes straight to the fixture.**

Example: Paragraph Organization by Spatial Pattern

When conditions permit, the normal way to install a new receptacle is to run a cable from a basement junction box to the point directly underneath the desired location, where it is brought up through the floor and inside the wall to the box opening. If there is no basement or the new convenience outlet is wanted on the second floor, the procedure is reversed to bring a cable down from the attic. Where no junction box is handy, a ceiling or wall lighting outlet or another receptacle can be the starting point. The job is actually simpler than it sounds, although some particular situations do create complications.

Before drilling any holes, make certain the circuit you plan to tap is

---

*From Norman Rothschild, "First Look," *Popular Photography*, May, 1970. Reproduced with permission.

**From *The Popular Mechanics Home Book of Electrical Wiring and Repair* by H.P. Strand. Copyright 1960; 1964 by H.P. Strand.

adequate. If it already has as many as ten outlets, you would be wise to pick another one. In the event the new receptacle is to be in the kitchen, dining room, breakfast room, pantry or laundry, install it on a special appliance circuit wired with No. 12 cable to conform with the Code. In other rooms, it can be connected to any general purpose circuit. For uniformity, plan to place the outlet at the same height from the floor as the older receptacles in the room.†

Inductive order. The inductive order or the method of induction could be discussed under several different headings in a book on technical report writing. Since inductive logic offers a basic useful pattern for organizing report materials, it will be discussed here, and elsewhere as necessary.

Persons trained in the application of the sciences know that induction is a method of reasoning a conclusion from observable facts. The mind moves systematically from observation of a set of particular facts to a general conclusion based upon them. Science and technology have a fundamental dependence upon the inductive method that is well known.

A fact can be verified—it can be proved again by demonstration. While interpretation of given facts varies and conclusions from them differ, the facts themselves can be reconfirmed and resubstantiated. When a writer chooses to organize his materials inductively, it is assumed that his facts are not merely assertions that cannot be proved. Conclusions are, as we have said, things about which reasonable people may agree or disagree, but facts are facts.

Organization by inductive order springs naturally from the writer's procedure in gathering his facts. Whether the particular facts are to be set forth chronologically or spatially depends on the nature of the subject, for they are arranged in whatever order will make it easiest to draw a logical conclusion from them.

Induction is not a fool-proof process. The report arranged inductively and the conclusion it draws can only be as convincing as the logic upon which they are based. Induction aims at generalization. It attempts to formulate general rules and con-

---

†From *The Popular Mechanics Home Book of Electrical Wiring and Repair* by H.P. Strand. Copyright 1960; 1964 by H.P. Strand.

clusions from particular data systematically collected and arranged. The problem is that faulty generalizations can be made and incorrect conclusions drawn from particular data.

There are three tests for avoiding faulty generalizations arrived at inductively. The writer asks himself if the generalizations are drawn from 1) single or isolated or too few particulars; 2) selected particulars; or 3) ignorance or prejudice. These general tests give rise to a check-list of questions which any report reader may ask himself, and which every writer should therefore know.

  A. Has the investigation covered a wide enough field?

  B. Are the conditions observed typical of general conditions, or are they special conditions prevailing only in the sphere of the investigation?

  C. Is the conclusion reached one that could reasonably be supposed to exist?

Phrased somewhat differently these questions may be posed thus:

  A. What are the relative number of the unobserved instances or facts?

  B. Do the instances observed or facts developed form a fair and sufficient sample? Are no exceptions discoverable?

  C. What is the degree of probability of the validity of the generalization or conclusion offered? Is the conclusion a reasonable or likely one?

Seen in this way, the inductive order of arranging material is both an outgrowth of and an aid to the scientific method of gathering facts for an investigative report. The process of bringing together facts leading to a conclusion provides a foundation for organizing the report.

Example: Paragraph Organization by Inductive Pattern

Amplitude Modulation

> The business of a radio station is to transmit sound, in some form, by electrical means. The first step in this process is to convert the sound into an electrical signal. This is done by a *microphone*, which transforms the sound waves into correspondingly varying voltage, and by an *audio amplifier* which strengthens that voltage. If both the microphone and amplifier are high-fidelity devices, the electrical wave-form put out

by the amplifier will be undistorted—that is, it will correspond very closely to the sound waves creating it.

Some means must be used to transmit this audio signal to the antennas of home radio receivers. Any attempt to transmit the audio signal directly would be doomed to failure under ordinary conditions. An average audio signal has a frequency of 5,000 cycles per second. To transmit a signal of this frequency efficiently would require an antenna twenty miles high! The next best method is to impress the audio signal on a second signal of much higher frequency known as the *carrier*. With its high frequency, the carrier can be efficiently transmitted using much smaller components.

One method of impressing the audio signal on the carrier is that known as *amplitude modulation*. In this method, the peaks of the carrier wave are elongated or shortened in accordance with the variations of the sound to be transmitted. If an intense sound should strike the microphone in the studio, there will be a marked difference between the heights of the elongated and shortened carrier wave peaks; if the sound at the microphone is weak, the peaks of the carrier wave will be more nearly equal. The word "peak" is the common equivalent of the engineering term "amplitude"; modulation is the act of changing. Hence, amplitude modulation is the act of changing the peaks of the carrier signal in the same way that the audio signal varies.

Pictorially, the audio signal can be represented by a continuous curve touching the peaks of the carrier wave. The negative peaks of the carrier are modulated in the same way as the positive peaks. Consequently, a representation of the audio signal by one curve tangent to the positive peaks of the carrier, and another curve tangent to the negative peaks, will show *two* audio signals, one the mirror image of the other. These two audio signals are completely opposite in phase, one of them rising at the same moment the other is falling.

This is the way the carrier arrives at the antenna of the AM radio receiver. The audio signal is the meaningful signal; it represents the sound the listener wants to hear. If the detector circuit in the receiver extracts both audio signals from the modulated carrier, the two signals will oppose each other, and nothing will be heard from the receiver's loudspeaker. It is then necessary to make use of only the positive half-cycles of the modulated carrier and to ignore the negative half-cycles. This operation, as we have seen in the preceding chapter, is one which the diode is suited to perform.*

---

*Reprinted with permission of The Macmillan Company from *Electronics* by Jesse Dilson. © by The Crowell–Collier Publishing Company, 1962.

Example: Paragraph Organization by Inductive Pattern

How a fluorescent lamp works is shown in Fig. 9, the diagram of a typical circuit. When power is turned on, current flows through the ballast to a pin contact attached to a heater, or electrode, in one end of the lamp. At first, this current moves on from the electrode via the other contact pin at the same end of the tube to the starter, which acts like a switch in closed position. From here, the current moves on to the other end of the tube, uses the pin contacts to pass through the heater and completes the circuit.

This action causes a red-hot glow in the electrodes, which are coated with an oxide giving off free electrons when heated. The pre-heat stage ends in a second or two when the switch contacts in the starter open to break the circuit. At the same instant, inductance in the ballast produces a voltage surge causing the current to arc from one heater to the other, and thus creating a new route to complete the circuit.

The arc stream travels the length of the tube through vaporized mercury and argon gas, which creates ultra-violet light. This itself is invisible, but it activates the phosphors coated on the inside of the tube, causing it to glow with a bright light.*

T. E. Huxley was a nineteenth-century English biologist who also wrote widely on anatomy, physiology, and other fields of science. He is remembered today as a strong early supporter of Darwin's theory of evolution. His writing in the final decades of the last century helped to explain the implications of Darwin's theory and to introduce the idea of science to a wide, popular audience.

His essay on "The Method of Scientific Investigation," which we reproduce below, serves to summarize further the discussion of scientific methodology. But the style of his writing is a useful model for technical writers. Since Huxley's audience is general, his style is more familiar than one would expect to find in a technical paper for technical audiences. But his choice of simple words, his system of organization, and his method of analysis which offers definition, comparison, example, and cause-to-effect reasoning (discussed in the next chapter of this text) are models for any technologist.

---

* From *The Popular Mechanics Home Book of Electrical Wiring and Repair* by H.P. Strand. Copyright 1960; 1964 by H.P. Strand.

a supplementary view...

THE METHOD OF SCIENTIFIC INVESTIGATION*

By Thomas Henry Huxley

Thomas Henry Huxley, Darwiniana Essays *(New York: D. Appleton and Company, 1895)*. Reprinted by courtesy of Appleton-Century-Crofts.

The method of scientific investigation is nothing but the expression of the necessary mode of working of the human mind. It is simply the mode at which all phenomena are reasoned about, rendered precise and exact. There is no more difference, but there is just the same kind of difference, between the mental operations of a man of science and those of an ordinary person, as there is between the operations and methods of a baker or a butcher weighing out his goods in common scales and the operation of a chemist in performing a difficult and complex analysis by means of his balance and finely graduated weights. It is not that the action of the scales in the one case and the balance in the other differ in the principles of their construction or manner of working; but the beam of one is set on an infinitely finer axis than the other, and of course turns by the addition of a much smaller weight.

You will understand this better, perhaps, if I give you some familiar example. You have all heard it repeated, I dare say, that men of science work by means of induction and deduction, and that by the help of these operations, they, in a sort of sense, wring from Nature certain other things, which are called natural laws and causes, and that out of these, by some cunning skill of their own, they build up hypotheses and theories. And it is imagined by many that the operations of the common mind can be by no means compared with these processes, and that they have to be acquired by a sort of special apprenticeship to the craft. To hear all these large words, you would think that the mind of a man of science must be constituted differently from that of his fellow men; but if you will not be frightened by terms, you will discover that you are quite wrong, and that all these terrible apparatus are being used by yourselves every day and every hour of your lives.

There is a well-known incident in one of Molière's plays, where the author makes the hero express unbounded delight on being told that he has been talking prose during the whole of his life. In the same way, I trust that you will take comfort, and be delighted with yourselves, on the discovery that you have been acting on the principles of inductive and deductive philosophy during the same period. Probably there is not one here who has not in the course of the day had occasion to set in motion a complex train of reasoning, of the very same kind, though differing of course in degree, as that which a scientific man goes through in tracing the causes of natural phenomena.

A very trivial circumstance will serve to exemplify this. Suppose you go into a fruiterer's shop, wanting an apple—you take one up, and, on biting, you find it is sour; you look at it, and see that it is hard, and green. You take up another one and that too is hard, green, and sour. The shop man offers you a third; but, before biting it, you examine it, and find that it is hard and green, and you immediately say that you will not have it, as it must be sour, like those that you have already tried.

Nothing can be more simple than that, you think; but if you will take the trouble to analyze and trace out into its logical elements what has been done by the mind, you will be greatly surprised. In the first place, you have performed the operation of induction. You found, that, in two experiences, hardness and greenness in apples went together with sourness. It was so in the first case, and it was confirmed by the second. True, it is a very small basis, but still it is enough to make an induction from; you generalize the facts, and you expect to find sourness in apples where you get hardness and greenness. You found upon that a general law, that all hard and green apples are sour; and that, so far as it goes, is a perfect induction. Well, having got your natural law in this way, when you are offered another apple which you find is hard and green, you say, "All hard and green apples are sour; this apple is hard and green, therefore this apple is sour." That train of reasoning is what logicians call a syllogism and has all its various parts and terms—its major premise, its minor premise, and its conclusion. And, by the help of further reasoning, which, if drawn out, would have to be exhibited in two or three other syllogisms, you arrive at your final determination: "I will not have that apple." So that, you see, you have, in the first place, established a law by induction, and upon that you have founded a deduction and reasoned out the special conclusion of the particular case. Well now, suppose, having got your law, that at some time afterwards, you are discussing the qualities of apples with a friend; you will say to him, "It is a very curious thing—but I find that all hard and green apples are sour!" Your friend says to you, "But how do you know that?" You at once reply, "Oh, because I have tried them over and over again and have always found them to be so." Well, if we were talking science instead of common sense, we should call that an experimental verification. And, if still opposed, you go further and say, "I have heard from the people in Somersetshire and Devonshire, where a large number of apples are grown, that they have observed the same thing. It is also found to be the case in Normandy, and in North America. In short, I find it to be the universal experience of mankind wherever attention has been directed to the subject." Whereupon your friend, unless he is a very unreasonable man, agrees with you and is convinced that you are quite right in the conclusion you have drawn. He believes, although perhaps he does not know he believes it, that the more extensive verifications are—that the more frequently experiments have been made and results of the same kind arrived at—that the more varied the conditions under which the same results are attained, the more certain is the ultimate conclusion, and he disputes the question no further. He sees that the experiment has been tried under all sorts of conditions, as to time, place, and people, with the same result; and he says with you, therefore, that the law you have laid down must be a good one, and he must believe it.

In science we do the same thing; the philosopher exercises precisely the same faculties, though in a much more delicate manner. In scientific inquiry it becomes a matter of duty to expose a supposed law to every possible kind of verification and to take care, moreover, that this is done intentionally and not left to a mere accident, as in the case of the apples. And in science, as in common life, our confidence in a law is in exact proportion to the absence of variation in the result of our experimental verifications. For instance, if you let go your grasp of an article you may have in your hand, it will immediately fall to the ground. That is a very common verification of one of the best established laws of nature—that of gravitation. The method by which men of science established the existence of that law is exactly the same as that by which we have established the trivial proposition about the sourness of hard and green apples. But we believe it in such an extensive, thorough, and unhesitating manner because the universal experience of mankind verifies it, and we can verify it ourselves at any time; and that is the strongest possible foundation on which any natural law can rest.

Chapter 7
THE JOB TECHNICAL WRITING DOES (CONT.)

Ways of organizing technical material for presentation and ways of developing it are so closely connected that it is difficult to separate them. Organization refers to the structure—the process of organizing, the pattern of the overall design. Development refers to unfolding more completely, to making something more available or usable by expansion of detail. In these senses, organization has to do with the exterior outline of the written material and development with the interior growth that fills in the outline.

But certain organizational patterns seem quite naturally to call for certain developmental techniques. Chronologic order, for example, lends itself to classification and division, to cause and effect, and to process or function as means of development. Spatial organization works effectively with definition and classification. Inductive order is developed chiefly by cause and effect. Yet, these distinctions are helpful only as teaching and learning aids. Rarely in the everyday practice of technical writing does one encounter simple examples of *this* pure pattern of organization or *that* pure method of development. Study of organizing and developing methods is of value only if the technician comes to recognize the range of possibilities open to him and then adapts them freely but logically to the needs of specific writing projects.

DEVELOPING THE EXPLANATION

By long practice, writers of explanation (or exposition, as it is also called) employ six basic methods for developing their material: 1) definition; 2) comparison and contrast; 3) example; 4) classification and division; 5) cause and effect; and 6) process or function. We shall discuss comparison and contrast and example as part of definition and include discussion of objective description as well, since all together they contribute to an extended definition.

Definition. Definition is a cornerstone of logical thinking. Since the purpose of definition is to distinguish what a thing is from what it is not, the technical writer must frequently employ it. Technical writing, as all scientific and objective writing, depends so much upon definition that one might say that defining is one of the prime functions of the technologist-writer.

Definition is the logical process by which the meaning of a term is stated. It is the process of stating and limiting the meaning of an object or concept.

A logical definition in the form of a sentence consists of three parts: the term, or the thing to be defined; the genus to which the term belongs; and the species to which the term belongs. Genus or class is any group of objects or ideas of which all the individual members have at least one characteristic in common. The function of the genus is to include the term. Species is a category of classification lower than a genus. The function of the species is to distinguish the term to be defined from other members of the genus—to exclude all members of the genus except the term defined.

To make a logical definition: 1) Name the term to be defined, 2) place it in an appropriate genus or class, and 3) place it in a suitable species, to differentiate it from other members of the same genus.

Technology is applied science.

A wheel is a circular body capable of turning on a central axis.

Aeronautics is the science that treats of the operation of aircraft.

An aeronautic technician is one versed in the technicalities of aeronautics.

In these examples of one-sentence logical definitions the terms to be defined are: *technology, wheel, aeronautics,* and *aeronautic technician*; the genera are *science, circular body, science,* and *one versed in technicalities*. That which remains in the sentences is the differentiation, serving to separate the terms from other species in the same genus.

In general, the beginning technical writer should make his definitions in the form of a complete grammatical sentence. Fragmentary sentences make for fragmentary definitions. For example: "wheel—a circular device that turns" is a fragmentary definition. The technical writer should strive to state his definition in simple and familiar terms. A definition that includes any word which in itself requires definition before the term can be understood is not a good definition. Also, definitions should be as brief as is consistent with clarity.

Definition employs words to tell us what a thing is. The reader of a technical report does not see on the printed page the *thing* the writer is trying to define (even with the help of visual aids). He sees only *words* the writer has used to represent the thing being defined. What the words *refer* to are the real thing, the true meaning. But the relationship between the words and what they refer to exists in the minds of the writer and reader.

In making a definition the writer must strive to be as specific as possible. It is more effective to define a carbohydrate as "a substance composed of carbon, hydrogen, and oxygen" than as "a neutral compound"; it is more useful to define an electrical charge as "the passage of a direct current" than merely as "a process of activating materials."

An adequate definition well distinguishes the thing being defined from all other objects in the same class. To define *gasoline* as "a propellant" is ineffective since it does not set off gasoline from all other propelling agents, liquid or gaseous. Similarly, to speak of *electricity* as "power efficiently and forcibly exerted" is unsatisfactory in that electricity is thus not distinguished from other types of energy.

The thing being defined and the genus into which it is placed must be presented in the same grammatical structure. One may correctly define: a halter is a rope or strap for leading or tying an animal; one may not say: a halter is made when an animal is led or tied by a rope or strap. One may correctly define: a nut is a perforated block with an internal, female screw thread; one may not say: a nut is useful on a bolt or screw for tightening.

Lastly, the technical writer should not use the word being defined or a derivative form of it in his definition. One cannot define *electrification* as "a common characteristic of everything that is electrified," or *automation* as "the state of being automated." Avoiding this weakness—and the other illogical errors noted above—will help the technician make definitions of real use to the reader.

Example: Paragraph Development by Definition

The theme of technical language, with which the preceding chapter was begun, might well carry into this one. Technical terms are meant to be precise, but what with the pressure of history and the loose talk of people who should know better, these terms begin to lose their precision and come to mean just about anything.

One of these terms is radio frequency, which we have been conveniently abbreviating r-f. Before the radio industry bloomed into its present eminence as the electronics industry, its domain was a narrow one and the language it used to describe that domain was correspondingly narrow. In those days, radio transmission and reception was more or less confined to the AM broadcast band, and so the term r-f referred to those frequencies ranging roughly from 600 to 1600 kc. Nowadays, however, what with the tremendous range of frequencies used by communications services of every type, the term r-f can mean just about anything from 600 kc on up.

The term "audio frequency" is more definite, since it is tied to sound; but even here there is some difficulty. It is hard to set an upper limit to the audio frequencies because the limit of hearing varies with different people. A minority—with youth on its side—can hear sounds of up to 16 kc (16,000 cycles per second) without straining. Possibly this is the upper limit. For the same reason, it is equally difficult to put a lower limit to the audio frequencies. Perhaps it is best to put the audio band in round figures and say that it is anything up to 20,000 cycles.

As long as the old trf receiver held sway, radio engineers spoke in terms of r-f and a-f. But with the coming of the superheterodyne, they were faced with the problem of giving the signal at the output of the superheterodyne's first detector a name. They solved that problem by calling the signal "intermediate frequency"—that is, a signal with a frequency lower than r-f but higher than a-f. And "intermediate frequency" it remains to the present day. As we have seen, the i-f can be just about anything, depending on the system in which it is found.*

Example: Paragraph Development by Definition

The basic elements of a fluorescent light circuit comprise the lamp, a starter and a ballast. *Lamps* are glass tubes of varying diameters and lengths which consume from six to 100 watts of power. The *starter* is a type of automatic switch control for the circuit. It may be a separate component installed in a starter socket or be included as part of the fixture switch. The *ballast* is a wire coil wrapped around an iron core to restrict power flow. Some fixtures have instant-start ballasts which do away with separate starters.**

Extended definition: development by example and by comparison-contrast. Depending upon the complexity of the term and upon the nature of the readership, a simple logical one-sentence definition may not be sufficient to fulfill the technical writer's purpose. In this case the writer will extend or expand his definition with greater detail, in as many paragraphs as are needed. He uses the one-sentence definition merely as a point of departure.

Methods of extending a definition are varied. The most common way to clarify a term by extension is to provide a specific example or instance. A new idea can be made clearer if a definite, concrete example or instance is given. In the definition of technology as applied science, for example, the writer might wish to offer some examples of applied science in several different fields. Examples offer excellent means of developing explanations.

Another means of extending a definition is by comparison

---

*Reprinted with permission of The Macmillan Company from *Electronics* by Jesse Dilson. © by The Crowell–Collier Publishing Company, 1962.
**From *The Popular Mechanics Home Book of Electrical Wiring and Repair* by H.P. Strand. Copyright 1960; 1964 by H.P. Strand.

and contrast. The thing to be defined is shown to be like or unlike something the reader is likely to know already. To make a comparison more effective, the thing to be compared should belong to the same genus, and the points of the comparison should be specific and precise. In the attempt to clarify the definition of technology as applied science, for example, the writer might like to choose an object such as a self-propelling gasoline-powered lawn mower. He could then compare and contrast the interests of a theoretical scientist in the object with those of a mechanical technician. While the pure scientist might be chiefly interested in the theoretical bases of propulsion, the mechanical technician would be interested in the technicalities of manufacturing, operating, or maintaining the lawn mower. Both theoretical and applied scientists are of the same genus. The contrasts between the points of interest each might have in the lawn mower are specific enough to extend the definition of technology as applied science.

Analogy is somewhat like comparison and contrast in extending a definition. In comparison, both objects belong to the same genus or class, as with pure and applied science above. In analogy the comparison is between one object in one class and another object in a totally different class. One might explain the flow of materials on an assembly line belt with the flow of water in a conduit, for example. The unfamiliar is thus made understandable by means of the familiar.

Other techniques employed as a means of extending a logical definition include the citing of negative details; restatement and reiteration of the logical one-sentence definition; and a combination of all these methods. With negative details the writer indicates what the thing to be defined is not. After clearing away any incorrect ideas, he proceeds to define what the thing is. Reiteration and restatement involve restating the definition in several different ways so that it may become clearer to the reader. There is also the method of employing several of these expository techniques in whatever combinations seem most appropriate for definition.

Example: Paragraph Development by Extended Definition

PPI Scope

    A common variety of radar equipment offers the PPI or plan position indicator display, in which a map of the territory surrounding the radar antenna appears on the face of the scope.

    The antenna of the PPI radar rotates continuously through a complete circle. As it does so, it emits rapid bursts of high frequency signals which cover a narrow sector of the scanned area. The time interval between the emission of the "main bang" from the antenna and the return of echo pulses is so very short compared with the speed of rotation of the antenna—the signal and its echoes travel with the speed of light—that the antenna can be considered stationary in that interval.

    While the antenna is in that relatively stationary position, it receives echoes in the form of a train of pulses. Since objects closest to the antenna reflect the r-f energy directed at them earlier than objects further away, the initial pulses in the train represent solid bodies nearest the antenna. The remainder of the pulses are spaced out in proportion to the remoteness of the objects reflecting them. This pulse train is applied as a video signal to the control grid of the PPI scope.

    The center of the display on the circular face of the scope represents the position of the antenna. This is the point struck by the tube's electron beam in its resting position. However, the beam is allowed to rest only for a moment. The instant the antenna emits its main bang of r-f, the electron beam is swung radially outward from the center of the tube face. As it does so, it traces a visible line on the tube's fluorescent screen, a line composed of light and dark gradations corresponding to the pulses fed to the electron gun grid. If the radial deflection of the electron beam sweeps the beam outward from the tube face center at a constant rate, the lights and darks of the traced line will be spaced in the same way as the echo pulses. Thus that line, as seen on the scope screen, will be an accurate picture of the terrain illuminated by the antenna's r-f beam at that moment.

    When the electron beam has finished its scan out to the limit of the tube screen, a highly negative blanking pulse is put on the grid to blank out the beam, and the beam is quickly snapped back to its resting point at the screen center. By the time the antenna lets loose its next r-f bang, it has rotated some small angle around its axis. When the bang occurs, the electron beam leaves its resting place to begin the sweep of a second radial line, painting a light and dark picture of the echoes from the terrain as it goes; the angle between the second and first scanned lines on the tube face is exactly equal to the angle through which the antenna has turned. As the antenna continues its rotation, the electron beam continues its radial sweeps. Thus by the time the antenna has turned through one whole revolution, the electron beam has painted a map of the whole circle of the surrounding area.

Since the antenna takes several seconds to complete a revolution—a rather long interval as time is reckoned in electronics—the picture traced by the first few sweeps of the electron beam should not be allowed to fade before the rest of the radial lines in the circle have been swept out. If the material of which the tube's fluorescent screen is made has a long *persistence*, all parts of the PPI pattern will remain visible simultaneously. A long persistence screen is one in which the light produced fades out completely only after a long time has elapsed.*

Objective description. Description as a form of discourse is designed to give a mental image of a thing. The aim of the descriptive writer is to create a picture. But there are in fact two distinct kinds of description—objective and subjective.

Objective or technical description is factual and impersonal; subjective description is imaginative and emotional, involving the feelings of both writer and reader. It is objective description, of course, which is required in technical writing.

The technical writer should not mistakenly confuse definition with description. Definition is essentially concerned with meaning, while description is chiefly occupied with appearance. Yet there are instances when objective description proves helpful in the course of extended definition.

A technical writer, for example, might be explaining the function of an engine oil-bath-air-cleaner on a rotary lawn mower. He has defined the element, located it for the reader, and compared it with an oil filter on an automobile. He has even offered a sketch of it, to support his explanation. As an aid to explaining which of three wing nuts must be tightened in the event of breakdown, he might describe the wing nut in question

---

*Reprinted with permission of The Macmillan Company from *Electronics* by Jesse Dilson. © by The Crowell-Collier Publishing Company, 1962.

by saying, "Seen from above, this wing nut is that one closest to the drive belt, and it forms the top of a triangle. The jack screw bracket forms the bottom left and the washout plug forms the bottom right of the triangle."

Example: Paragraph Development by Description

A vertical gyro is a two-degree-of-freedom instrument whose gimbal displacements about each output axis constitute a measure of angular deviation from the local vertical axis. The rotor spin axis is maintained parallel to local gravity vertical by a gravity sensing device. The term *vertical gyro* derives from the fact that it is used to measure displacement from the vertical reference. A vertical gyro effectively serves the same purpose as a pendulum, with the advantage that a maneuver does not cause it to oscillate. This advantage is particularly desirable since attitude information is most needed during maneuvers.

A vertical gyro usually has 360 degrees of angular motion about the outer gimbal axis and ±85 degrees of angular motion about the inner gimbal axis. Restriction of motion about the inner gimbal axis is accomplished by mechanical stops which are necessary since ±90 degrees of displacement of the inner gimbal with respect to the outer gimbal causes "gimbal lock." Since roll motions of 360 degrees are more common than pitch motions of ±85 degrees, the conventional practice is to use the inner gimbal axis for measuring pitch and the outer gimbal for measuring roll. At the zero reference position, the spin axis, inner gimbal axis, and outer roll axis are mutually orthogonal to one another.*

Classification and division. Technical explanation employs classification and division with almost as much frequency as definition. In classification one identifies the general class to which a particular example belongs; in division one separates a general class into its various exemplary parts.

Classification is indispensable to systematic thought. It helps us keep our thinking clear and structured. The mind of the writer is given a direction. Order is achieved by dividing or separating things into logical groupings. Like definition, classification is essentially a form of analysis, that is, a way of dividing

---

*Singer—General Precision, Inc. Kearfott Division. Reprinted with permission.

things into related groupings to secure order between parts and the whole.

Classification deals with classes or plural subjects. Division deals with parts of a singular subject. Technologies may be classified, but mechanical drafting technology must be divided or broken down into its component parts for analysis. If technologies are classified, the results of the breakdown will still be technologies—refrigeration, textile, or diesel engine technology, for example. But if a single technology is divided, the result will not be a technology but a part or a phase of one—design, manufacture, operational—no one of which is a whole technology but is an integral portion thereof. Classification breaks things down into kinds of classes; division breaks things down into parts, steps, or stages.

In much analytical technical writing, classification and division necessarily work together. First of all, the writer decides on a basis of classification. Outboard motors can be classified by style, horsepower, weight, price, or kinds of propellors, for example. The writer establishes the basis of classification which best suits his objective. But once having done so, he must see to it that the same single basis of classification is used throughout the report.

If, for example, he is going to analyze outboard motor A on the basis of horsepower, he proceeds to define horsepower and to indicate that it is the basis of classification. He has already moved down the scale from all motors, to motors used on boats, to outboard motors used on boats, to outboard motors of X horsepower used on boats, to outboard motor A with X horsepower used on a boat. Once his report is underway, the writer will divide outboard motor A into those component parts relevant to the horsepower. He may not properly shift from his horsepower basis of classification to some other basis in any division of the report. He should not, for instance, discuss the piston and ring in terms of horsepower and then introduce price considerations to the section dealing with the carburetor. The exposition should properly proceed in parallel fashion on all levels of division of the report, in terms of the established basis of classification.

Example: Paragraph Development by Classification and Division

Amplifier Classes

In electronics, as in life generally, the advantages of a particular method are usually accompanied by certain disadvantages. The type of amplifier predominantly used in audio and r-f circuits is, as we have seen, useful because it produces an undistorted output; but, since plate current flows even with no signal applied to the grid, the circuit is inefficient. Such an amplifier is termed *Class A*. According to the established definition, a class A amplifier is one in which the grid bias and signal voltages are such that plate current flows at all times.

There are other types of amplifer which are more efficient than the class A type, and whose tendencies to distortion are compensated for in one way or another. One such is the *Class B amplifier*. In this type, the bias voltage applied to the grid is approximately equal to the cutoff value. Hence very little plate current flows except when a signal voltage is applied to the grid; then, only the positive half-cycles of the a-c signal voltage are reproduced in the output of the amplifier.

In between the class A and class B variety is the *Class AB* amplifier. Plate current in this type flows for more than half the cycle of the input signal voltage but less than the entire cycle.

Finally, there is the *Class C* amplifier, in which the grid is biased so negatively as to be beyond cutoff. With no signal on the grid, definitely no plate current can flow; when a signal is put on grid, plate current flows for less than half the signal voltage cycle. Class C amplifiers are most efficient because plate current consists for the most part of powerful and useful pulses. Consequently, they are used a good deal in transmitters of all types.

One final note is needed to round out this system of amplifier classification: If grid current flows for any part of the input signal cycle, a figure 2 is written after the class letter; if no grid current flows, a figure 1 is used. This method of letter-and-number amplifier classification appears so frequently in electronic literature that it must find a place in any good book on the subject.*

Cause and Effect. Cause to effect and vice versa are what we spoke about in the preceding chapter as the inductive pattern of organization. The inductive method may be said to emphasize

---

*Reprinted with permission of The Macmillan Company from *Electronics* by Jesee Dilson. © by The Crowell-Collier Publishing Company, 1962.

the *how* of things—to speak of cause and effect is to stress the *why* of things. But except for discussion and clarification these aspects cannot really be separated.

One could very simply summarize the purposes of the inductive order by saying that in a report it serves to present facts, to formulate conclusions, and to depict relationships between causes and effects.

Once it has been determined how things generally happen in a certain way, that is, once a conclusion has been induced from a particular set of circumstances, the mind wishes to know why these effects follow from the causes. But though events occur together there is no certainty of a causal relation between them.

The reader always asks himself whether what the report calls the cause does adequately explain the effect. If there is an excessive oscillating effect when the wheel slows to 400 revolutions per minute, can the cause be properly attributed to a defective armature? The logical mind proceeds to ask if there are any other forces or factors that may be involved in producing the oscillation effect. The inductive order requires examination of possible other causes.

The logical and systematic inductive search from observable effect back to possible causes follows a course springing from the nature of the subject. As the investigation proceeds scientifically from cause to effect and back again the development of the technical report merely follows.

All methods of developing paragraphs are expressions of the natural workings of the mind. Any refined intelligence will informally seek to define things and ideas as it carries on the everyday business of life. The beginning technical writer would do well to recognize the natural parallels between the methods of expository development and the usual patterns of thinking. Technology, and therefore technical writing, is especially concerned with cause and effect. What brought this change about? What will happen when we do this? It is basic to the nature of science to seek the causes of an existing fact or follow its effects or consequences.

Cause and effect organization of material is essentially a technique for problem solving. One first recognizes the problem. It is best not to assume that it has only a single cause, for it may have any one or all of several causes. Understanding this, one examines various choices or possibilities for solving the problem. Each possibility will be carefully considered before a determination is made as to which is really the best.

Example: Paragraph Development by Cause and Effect

Compared with incandescent lighting, fluorescent lamps offer up to three times as much light per watt of power used. The lamps also have a much longer useful life and provide a softer, diffused light which is more comfortable to the eye. As a result, it has become the preferred type of lighting in kitchen-ceiling and under-cabinet fixtures, and its popularity in bathrooms, recreation rooms, basements and other work areas is steadily increasing. It plays an important part in such indirect lighting methods as valance lighting to feature ceilings or particular wall areas and room lighting through translucent ceiling panels.*

Example: Paragraph Development by Cause and Effect

Shortly after Thomas Edison invented his incandescent lamp, he performed an experiment with it which helped bring the electron to the attention of the world. Into the glass bulb of the lamp he sealed a metal plate wired in series with a battery and an ammeter. Then he turned on the current to light up the lamp filament, and observed some strange results: when the positive pole of the battery was connected to the metal plate inside the bulb, the ammeter showed that an electric current was flowing through the series circuit; but when the battery connections were reversed, the ammeter registered zero.

Reasoning from the basic electrical law that like charges repel and opposite charges attract, Edison was able to interpret clearly the results of his experiment. Current could flow through the circuit because electrified particles crossed the gap between the two electrodes in the lamp. These particles could only have been urged across the gap by the force of attraction of the plate when it was positive. They must therefore have been negatively charged. With the reversal of the battery connections, the plate in the bulb was given a negative potential; it then repelled the negative particles to halt their movement from the filament, and current ceased to flow.

---

*From *The Popular Mechanics Home Book of Electronical Wiring and Repair* by H.P. Strand. Copyright 1960; 1964 by H.P. Strand.

Later experimenters confirmed this explanation of the "Edison effect" and further discovered that these invisible particles of negative electricity—which they termed *electrons*—are so tiny as to have only 1/1800 the mass of the lightest atom, the atom of hydrogen.*

Process or Function. Process analysis is generally concerned with the *how* of things. Causal analysis, as we have seen, is more preoccupied with the *why* of things. Since a large portion of technical writing has for its aim the explanation of how things function (as well as why), the technologist will often find himself adapting his material to the process method of presentation.

In its most basic form, process analysis consists of systematically breaking down an object into its component parts and explaining the function of each. The system of developing the material will spring from the object itself and from the purpose of the explanation. It is correct to think of the process pattern of paragraph development as an offshoot of the spatial pattern of organization. If the writer were, for example, explaining how a water purifying device functions, he would probably choose to unfold the process with respect to the inflow and outflow of the water. In this manner the dispersing mechanism which distributes the incoming water throughout the device would be the first step of the process to be explained. Next would probably come an explanation of the function of the polyethylene prefiltering device. Explanation of the process by which the activated carbon, enclosed in a plastic container, actually filters the water, might then follow. Before explaining the outflow device, the writer would probably choose to explain the mechanism which assures the bottom-to-top water flow. Along with this could come the exposition of the water anti-channeling process.

Though other systems for organizing the explanation of the total process are possible, the audience for the report would ultimately dictate the choice. The foregoing method was adapted to the needs of the individual homeowner who might

---

* Reprinted with permission of The Macmillan Company from *Electronics* by Jesse Dilson. © by The Crowell–Collier Publishing Company, 1962.

require the explanation as a prelude to installing the device himself. The installation process, of course, would be the subject of a separate process-explanation. As seen in the preceding chapter on organizing material for explanation, the spatial pattern allows discussion of a mechanism according to function, physical characteristics, or parts. Since function or process is important in technical writing, we draw attention to it here again as a means of developing material once it is organized.

Example: Paragraph Development by Process

The process we are talking about (there are others that could be called stabilization) has two main principles.

The first one incorporates developing agents in the paper. So long as the paper is dry, the agents remain inert. But when the paper is wet and especially when the paper is wet from the strong alkaline solution you put in the stabilization processor, the developing agents are activated. This eliminates virtually the whole "induction period"—the initial interval in ordinary developing when the print is submerged but nothing is happening. Another result is that development is limited to a predetermined extent. The preset development is, on the whole, desirable. It is helpful in that it tends to overcome problems such as varying temperature of processing solutions, wavering motor speed, and increased chemical concentration in solutions because of water evaporation. It is a hindrance in that it permits almost no error in exposure, a matter we'll get deeper into a little later.

The second main principle that made stabilization possible is leaving the fixer in the emulsion. Photographers who have had drilled into them the utter importance of washing out the last trace of fixer may find it hard to believe this is Right Thinking, but it is the basis of stabilization. As long ago as 1893, it was found that leaving a lot of fixer to dry in the print would do a fairly good job of making it last. Investigations after World War II showed that no attempt to wash the fixer out of the print at all was superior to a mediocre attempt. It was out of this research that modern stabilization processing was born.*

---

*From "Stabilization Processers," by Wallace Hanson. *Popular Photography*, May, 1970. Reproduced with permission.

Chapter 8
PARTS OF THE REPORT

A technical report, as any report, consists of three general parts or sections. Each of the three general parts may consist of several different subsections of subsidiary units labeled separately to serve the special needs of the particular writer and reader. But the three general divisions of the report have distinct purposes regardless of the number of subunits of which each may be comprised and variously labeled.

The three broad parts of the technical report are the introduction, the body, and the conclusion. It is correct to think of these as beginning, middle, and end.

INTRODUCTION

The introductory or beginning part of the report states the subject or purpose of the writing and prepares the reader for what will follow. The reader is presented with sufficient background to the subject to allow the main points which will follow in the body of the report to stand out clearly. The purposes, scope, objectives, or limitations of the study may be presented. Or a brief historical survey of the subject may be given, to place the report in perspective. If certain theories or principles, or general information is required for understanding the nature of the technology, this will be given in the introduction section.

Some writers will prefer, once they have stated their purpose and objective, to give a short review of the chief methods followed in the study. Others will refer to their sources of information or their general methods of procedure, if these seem of special interest or importance. In fact, the writer of a technical report is left free (to a degree) to make his introductory section as long or as short as he thinks necessary to prepare the reader most effectively for the body of information to follow. In practice, the introductory section ranges in length from one paragraph to several pages; rarely is it longer.

The general heading used for the opening section is often simply Introduction. In place of this, or as a subheading, some writers will use Statement of Purpose, or Statement of Objective, History of the Problem, or Background to the Report, depending on what they choose most to emphasize. Other headings generally used for the opening section are Aims and Objectives, General Plans and Procedures, Reasons for the Report. No matter which specific heading seems most useful to the individual writer, the purpose of the opening section remains the same: briefly and clearly to tell the reader what the report is all about. If the introductory section does not satisfy this aim, according to the specific points of emphasis in the body of the report, then the introductory part is a failure.

Example: Introductory Section of a Technical Report

GENERAL

The term "synchro" is a generic one covering a range of AC electromechanical devices which are used in data transmission and computing systems. A synchro provides mechanical indication of its shaft position as the result of an electrical input or an electrical output which represents some function of the angular displacement of its shaft. Such components are basically variable transformers. As the rotor of a synchro rotates it causes a change in synchro voltage outputs.

Major types or classes of synchros described here include torque synchros, control synchros, resolvers, and induction potentiometers (linear synchro transmitters). Each class is treated separately and in detail, but before describing the particulars pertaining to each type, properties and characteristics common to virtually all types of synchros are considered. Although each class of synchro is used in different applications, their construction and the physical laws governing their performance are similar, thus making it possible to apply two basic schemes in defining their function. The first is the *transfer function*, which is a relationship between output response to all inputs. In a synchro receiver, input is an AC voltage while output is mechanical rotation or torque. With this exception, input for all other units is mechanical shaft rotation, with output expressed as an AC voltage. When static, or at slow shaft speeds, synchros are essentially transformers with variable coupling coefficients. In applications requiring higher rotational speeds, the transfer function is complicated by speed-voltage terms. That is, part of the secondary voltage results from cutting an alternating magnetic field by moving conductors. Further, when synchros are used in cascade, the overall transfer function is not the product of all transfer functions in the loop, because loading effects significantly alter each function. In such circumstances, another approach is employed, and is termed the *equivalent electrical circuit*. This differs from the transfer function in that it relates input and output electrical quantities rather than electrical (voltage) and mechanical (torque) factors. The use of an equivalent electrical circuit, therefore, permits calculation of loading on adjacent electrical devices, and the parameters involved are measurable at unit terminals under unloaded conditions. This method, then, provides results that are more meaningful for the applications in which it is used. The choice of either approach—transfer function or equivalent circuit—depends on their compatibility with varying environmental and operating conditions.

There are, of course, many other synchro properties or characteristics which must be considered by the designer when planning a system for particular application. Those which are generally applicable to virtually all types of synchros are described in this section, but details relating to these and other characteristics will be found under sections which deal specifically with the particular type of synchro to which they apply.*

The preceding introductory section is taken from a booklet that serves as a technical information bulletin for engineers who may use the company products. There are five main parts to the bulletin, covering Motors, Motor Generators, Synchros and Resolvers, Electronics, and Servos. Each main part of the bulletin has its general introduction. The example given here is the opening section of Part III of the bulletin, the section entitled Synchros and Resolvers.

The writer directly states the purpose of his writing by providing a definition of the terms to be covered in this one section

---

*Singer–General Precision, Inc. Kearfott Division. Reprinted with permission.

of the report. He offers a broad general view of the subject before breaking it down into those particular aspects which he plans to treat. This general introduction not only defines the term with which the section of the report will deal, but it offers a preview of the system of classification and division that will provide the plan around which this part of the report will be organized. The reader is adequately introduced to the subject and offered an outline of scope, objectives, and organization of the section.

Example: Introductory Section of a Technical Report

> IN BRIEF: *The tactical posture of our air forces—including ground support, fleet air defense, and air superiority—must be considerably updated to meet the challenges of the seventies. Perhaps the most pressing need is the potential for mastering the enemy's air space. To carry the fight to the enemy, the next generation of fighters to do that job will have to have lots of thrust. And since speed doesn't win close-in maneuvering contests, agility will have to be the fighter's forte.—P.H.*

Tactical airplane missions today are not too different from what they were 20 years ago. But the demands on aircraft are. Today aircraft must fly faster, further, and higher. A tactical plane may have to thunder at supersonic speeds at altitudes beyond 50,000 ft to intercept an intruder while the intruder is still well away from friendly territory. The same plane may also have to be slow enough to settle in on the short deck of an aircraft carrier. Or a single aircraft might be called upon to be a flying bastion, bristling with air-to-air missiles, or an agile interloper equipped to tangle with the enemy in close-in combat.

The airplane designer would like nothing better than to create a special aircraft for each mission. However, purse strings tie up the designer's freedom even as mission requirements grow more difficult to meet.

The problem is one that tactical aircraft designers will be struggling with in the coming months, for the Department of Defense is preparing to make some decisions about the next generation of fighter aircraft. We would like to show in some detail how the problem breaks down, to examine the methods for analyzing the problem, and to show what all this leads to in the way of a tactical airplane design for the seventies.*

---

*From Joseph Rees and Arnold Whitaker, "The Next Generation of Fighter Aircraft," *Science and Technology* (October, 1968).

The preceding introductory section is not labeled or titled. The report appears in a journal for scientists and technologists, and it provides—a common practice—a brief or abstract of the scope of the report. Since the title of the report, "The Next Generation of Fighter Aircraft," is printed just above the brief, there is no need to provide a subtitle for the introductory section. It is clear from the content that the opening paragraphs are introductory. The section ends simply by stating what the report that follows will do: "show in some detail how the problem breaks down, . . . examine the methods for analyzing the problem, and . . . show what all this leads to . . . ."

Example: Introductory Section of a Technical Report

> This report will describe a method of improving weight control by improving the sizing and operation of filling flasks which are frequently one of the most common, yet least recognized, causes of off-weight packages due to head to head variations.
>
> While it is generally true that the flasks are machined to fine tolerances and when adjusted should all expand or contract an equal amount, it is recognized from a practical standpoint, however, that the flasks often do get out of adjustment. This can occur due to a short-term buildup of product such as sugar in the filling flask, improper sizing or adjustment when new flasks are installed, damage occurring during cleanouts or changeovers, etc. It is then likely that several flasks will be affected and may, consequently, produce underweight or excessively overweight packages.

The foregoing introduction is from a report designed for engineers and technologists concerned with problems of packaging various products. The opening one-sentence paragraph states the purpose and scope of the writing and prepares the reader for what will follow. The second paragraph defines the problem and directs attention to the point of emphasis with which the report will concern itself.

Current trends in technical report writing are away from lengthy, extended introductory sections. In fact, some of the best contemporary technical reports eliminate the introductory section entirely. Many writing experts believe that placing the introduction in the prominent first position tends to divert attention away from what may be uppermost in the reader's mind, the conclusions and recommendations. Variations of the conventional order of presenting report materials will be discussed at the end of this chapter.

BODY

The heart of the report is what makes the report unique. The body contains all those specific details which, added together, make the report distinctively one thing and not another. The reader is told clearly what methods are followed and which materials are used. He is also told what results have occurred with these materials and methods.

Materials are identified with specific details. Specifications are listed, basic sizes and shapes are identified, and all of the particular characteristics and attributes of the material necessary to recognize it unmistakably are cited so that there may be no confusion. The body of the report informs the reader about those precise substances that the report deals with, states what methods have been followed in using these particular things, and indicates what results have occurred.

The body section does not generally concern itself with analysis or interpretation of facts. All the steps in the process are presented in logical sequence or order. The reader is given as much or as little detail as is needed for him to comprehend fully what is explained. By the time the reader completes the body of the report he should know definitely the *what* and *how* of the subject; the *why* is reserved for the end part of the report, beyond the body.

From the examples which follow of the body section from published technical reports, it may readily be observed that considerable flexibility is exercised in labeling various subsections. Each body section of a report presents its own special requirements, and the writer is always free to organize and present his central materials in the mode which seems most logical and clear. All of the following examples are presented without the graphic materials which accompanied the original reports. Nonetheless, even without the visual aids referred to frequently in the bodies, the student may observe that the body of the report does appropriately cover materials, methods, and results of the study, though a variety of different labels and headings are employed in practice.

Examples: Body Sections of Technical Reports

Example A

Bill of Materials

4—5" woofers (Olson No. S-845)*
4—2-3/8" horn tweeters (Olson No. S-846)*
1—Two-way crossover network (Olson No. HF-102)*
1 pkg.—Acoustical fiberglass (Olson No. HF-17)*
2—24" x 12" pieces of ¾" fir plywood for sides
2—24" x 10½" pieces of ¾" fir plywood for sides
2—14" x 14" pieces of ¾" hardwood plywood for top and bottom
4—13" x 1-5/8" pieces of ¾" hardwood plywood for foot pieces (miter cut ends to 45°)
1—9' length of ¾" x ½" trim for top and bottom (see text)
1—8' length of ½" outside corner hardwood molding for corner trim
1—144" length of ¾" x 1/48" wood veneer (Shurwood wood tape or similar) for plywood edges
4—10½" length of 1" x 2" pine for top and bottom cleats
4—7¼" lengths of 1" x 2" pine for top and bottom cleats
Six-penny finishing nails for attaching sides
Three-penny finishing nails for attaching trim
32—#8 x ¾" panhead sheet metal screws for mounting speakers
8—#8 x 1¼" flathead wood screws for attaching top
8—#10 x 2" flathead wood screws for attaching foot pieces
Misc.—Grille cloth (see text); wood glue; flat black paint; stain, sandpaper; wire; solder; etc.
*Olson Electrics, Inc., 260 S. Forge St., Akron, Ohio 44308

Construction. The enclosure can be built with common hand tools, though 45° miter cuts for the "foot" pieces and trim will improve the appearance. Cut out the parts to the dimensions shown in Fig. 1. In addition to the speaker cutouts. Drill two guide holes for screws through each side piece about 3/8" from the top edge and 5" apart. Glue and nail together the sides to form the column as in Fig. 2. Then coat the exterior surfaces of the column with a flat black paint.

Prepare the 14" square top and bottom pieces. Use the template supplied with the crossover network and a piece of carbon paper to make the cutout for the crossover on the bottom (see Fig. 3). Remove the cutout with a sabre or keyhole saw. Center the top and bottom on the open-ended column and outline the position of the sides against the end plates with a pencil.

Attach 1" X 2" cleats with glue and #8 X 1¼" flathead wood screws on the interior surfaces of the top and bottom plates as shown in Fig. 4. The cleats should fit within the space outlined by the pencil marks to allow screws to be driven through the enclosure sides into the cleats.

Next, cover the plywood edges of the top and the bottom with wood veneer edging to match the veneer on the plywood. Use a razor blade to cut a piece of ribbon veneer slightly longer than the panel. Coat the plywood edge and the rear surface of the veneer with contact cement. Allow the cement to dry for 10 to 20 minutes until it is tacky but does not stick to your finger. Then apply the veneer, but don't let the surfaces touch until the veneer is in exact position. The cement will adhere on contact; but to make sure the entire surface is tightly bonded, place a small block of wood against the veneer and tap with a hammer. Move the block and tap it along the entire length of the veneer. With a razor blade, trim the ends of the veneer to the proper length. Then sand the edges, using a small wood block covered with fine (4/0) sandpaper, slightly rolling the top edge to blend the grain of the veneer with that of the plywood.

Coat all matching surfaces between the bottom cleats, the bottom plate, and the bottom edges of the column with wood glue. Attach the bottom by driving nails through it into the lower edges of the four sides. If you have a good fit between the parts, the glue will be sufficient for proper sealing. If not, add screws through the sides into the cleats. Then check for air leaks and caulk the corner joints if necessary.

Install the four miter-edged foot pieces on the bottom plate with glue and eight #8 X 2" flathead wood screws. Feed the wires from the crossover network into the enclosure and install the network on the bottom, using the ten screws supplied with it.

Now mount the woofers with #8 X ¾" panhead screws. Locate the positive terminal of each woofer (may be identified by a red insulating washer between the terminal and the speaker frame; negative terminal has white washer). Wire the woofers according to Fig. 5. Then check the polarity of the system by connecting a flashlight battery to the crossover terminals. For proper phasing, all woofer cones should move together in one direction, either outward or inward.

Next, mount the tweeters with panhead screws; wire them according to Fig. 5; and follow the instructions supplied with the network to complete the speaker hookup. Connect the system to an amplifier and check the operation of the tweeter control; clockwise rotation should increase the sound level of the tweeters.

Fill the enclosure with loose fiberglass. One 72" X 18" sheet of Olson fiberglass is the minimum amount that should be used. Cut the batting into pieces about 18" X 10½", and insert them through the openings at the corners of the enclosure to fill the lower part up to the woofers. Then cut smaller pieces, about 3" X 10", to fit in the space between the woofers. The level of the fiberglass should extend to the level of the tweeters.

Set the top in position, and mark the correct positions for screws on the inside cleats. Remove the top and drill 1/8" guide holes in the cleats. Cement a thin gasket of polyfoam or felt along the top edges of the sides. Then replace the top and anchor it with screws driven through the sides and into the cleats as in Fig. 6. The screws will be in the proper position to draw downward on the top, compressing the gasket. If necessary, weight the top to bring the guide holes in line with the screws.

A piece of grille cloth 2' X 4' will fit the enclosure column, wrap-around style. However, if the grille cloth you select does not have a strong vertical or horizontal pattern, you might be able to economize by cutting a 1'-wide strip from one end of a square yard of cloth. Use this strip in a vertical position and wrap the 2' X 3' remaining strip around the enclosure. However you plan it, measure the distance around the enclosure before you buy the cloth or order an extra few inches to allow for mistakes.

Fasten the cloth at one corner with tacks or staples. Stretch the cloth across each side, and add a few tacks or staples at each corner to hold it taut. The vertical wood strips will cover the corner staples.

The exact lengths of the top and bottom trim pieces will depend on the thickness of the grille cloth so they must be cut to fit. These pieces of trim can be made either from solid wood or plywood with veneer-covered edges. Use small finishing nails to attach them to the top and bottom of the enclosure.

Finally, cut outside corner molding to fit tightly between the top and bottom trim. Stain and finish this molding to match the other wood before attaching the pieces. (Other surfaces can be stained and finished in place.) When they are dry, attach the corner molding with small brads (see Fig. 7).

This completes the construction of your Omni-Eight. Connect the leads from your amplifier and give it a listening test. You may find that a change in position of the Omni-Eight in your listening room requires a different tweeter control setting.*

Example B

Parts List

BP1,BP2—Binding post (one red, one black)
C1—6-µF, 15-volt electrolytic capacitor
C2,C3—0.1-µF capacitor
C4,C5—500-pF, 20-kV "doorknob" capacitors (used in TV high voltage)
I1—12-volt pilot lamp (GE 1815 or similar)
Q1-2N173 or HEP 223 power transistor
R1—15-ohm, 5-watt resistor
R2—2.2-megohm, ½-watt resistor
T1—TV horizontal flyback transformer (Stancor HO-290 or similar)
V1,V2—1X2B high-voltage rectifier tube
Misc.—Heat sink (Wakefield NC621B or similar); insulated mounting hardware for transistor; silicone grease; pilot light holder; 9-pin anti-corona high-voltage rectifier sockets (2); ceramic supports for sockets (2); feedthrough insulator (E.F. Johnson 135-48 or similar); length of ½" high-voltage tubing; length of 1/8" plastic tubing; length of nichrome wire; suitable metal chassis.

Power Supply. The circuit for the high-voltage power supply is shown in Fig. 3. For safety, the device is enclosed in a grounded metal container and the high-voltage output is taken through a feedthrough insulator.

The supply is a simple transistor oscillator using extra windings on a conventional TV high-voltage horizontal output transformer. Two of these extra coils, a primary and a tickler feedback (*L1* and *L2* in Fig. 3), in conjunction with the transistor, form a regenerative feedback network similar to that used in receivers. When the power is turned on, current flows through *L1* and the transistor. The magnetic field set up by this current generates a voltage in *L2* that increases the forward bias on the emitter of *Q1*. The collector current through *L1* then increases. Eventually, the core of the transformer saturates and the magnetic field

---

*From "Omni-Eight Speaker System," by David B. Weems. *Popular Electronics*, February, 1970. Reproduced with permission.

around $L2$ stops building so that the emitter bias is reduced and the collector current drops. The process is then reversed. The magnetic field set up by the decreasing collector current produces a voltage in $L2$ that drives the transistor to cutoff. When no current flows through $L1$, there is no voltage across $L2$ and the emitter returns to ground potential. The cycle then repeats. The oscillator frequency is near the upper end of the audible range.

The transistor is biased by $R1$, which is bypassed by $C1$. Capacitors $C2$ and $C3$ protect the transistor from static discharges.

The current through $L1$ varies from zero to about 5 amperes. Because of the turns ratio between $L1$ and $L3$, about 5000 volts a.c. is developed across $L3$. A voltage-doubler/rectifier combination ($V1$ and $V2$ and $C4$ and $C5$) raises the voltage to about 10,000 volts d.c.

Caution. Although the current is low, voltages at the 10-kV level can be very dangerous. Do not, under any circumstances, turn on this high-voltage generator unless the case is completely closed and the high-voltage feedthrough is well in the clear. When the system is turned off for any reason, always discharge the high-voltage terminal using an insulated cable, with one end secured to ground and the other end held at the end of an insulated rod to touch the terminal.

Parts List

$C1$—5000-µF, 15-volt electrolytic capacitor
$D1,D2$—Silicon rectifier (GEA40B or similar)
$T1$—Power transformer, secondary 24 volts, center tapped
Misc.—TV "cheater" cord and receptacle; diode heat sink (Delta NC403K or similar); capacitor clamp; mounting hardware; etc.

Power Supply Construction. Remove the insulated filament winding from the flyback transformer. If there is a sponge-rubber pad between the core and the mounting bracket, remove it. Caution: the core is made of a brittle ferrite material in an epoxy binder. Therefore, do not force or twist it in any way. Gently remove the rectifier plate connector lead from the coil. Make sure that you can identify the high-voltage winding terminals.

Solder a two-lug terminal strip to each side of the mounting bracket as shown in Fig. 4. Wrap a layer of insulating tape on the bare horizontal ferrite core, feeding the tape between the core and the mounting bracket. Wind 12 turns of #18 stranded hookup wire in a close layer around the core. This forms $L1$. Solder the ends to the bottom insulated tie points of the terminal strips. If the winding does not pack tightly, remove it, and rewind with a slightly larger wire.

Wind a five-turn coil, using the same gauge wire, on top of $L1$. This forms $L2$. Solder the two ends to the top lugs on the terminal strips. Wind $L2$ in the same direction as $L1$ with the windings spaced evenly across $L1$.

Using the same type of wire originally used for the filament winding (removed in an earlier step), wind the two one-turn coils between the turns of $L2$. These form $L4$ and $L5$ and will be connected to the filaments of the rectifier tubes. Locate one end of the high-voltage winding and connect it to the nearest ground—the transformer mounting bracket will do.

Obtain a metal box, large enough to accommodate the transformer and the rectifier tubes, yet small enough to fit between the metal column and the side of the cabinet. It should be less than 10" high (including the insulator for the high-voltage feedthrough) so that the entire assembly will fit below the shelf in the cabinet.

The transistor is mounted on a heat sink using appropriate hardware and insulating material. Coat both sides of the transistor insulator with silicone heat-conducting grease. The heat sink assembly is mounted at the outside lower end of the rear panel to keep it away from corona discharges set up in the high-voltage section (see Fig. 5). Appropriate holes must be drilled in the rear panel to mount the heat sink and to provide access to the transistor terminals.

The flyback transformer is mounted at one side of the power supply enclosure (see Fig. 6) so that the high-voltage and filament leads face the two rectifier tubes. The tube sockets (of the anti-corona type) for the rectifiers are mounted on ceramic insulators, one on the top and the other on the bottom of the enclosure. Mount the other components as shown in Fig. 6 and wire them, point-to-point, as shown in Fig. 3. Take care to make neat, smooth joints and avoid sharp edges to prevent corona discharges. Resistor $R2$ connects from the rectifiers to the feedthrough insulator.

Checkout. Connect a 2- to 3-volt d.c. source to the battery input terminals, with positive to ground. A pair of flashlight D cells will do. The circuit will oscillate with this low supply but voltage levels will be down. Try to draw an r.f. arc from the transformer high-voltage terminal using a well-insulated screwdriver. If there is no arc, even a small one, reverse the connections to $L2$. In some cases, it may be necessary to add or remove turns from $L1$ to obtain the proper core saturation.

Once you know the oscillator is working, connect 12 volts d.c. to the circuit and, being extremely careful, measure the voltage at the filaments of the rectifiers. It should be between 1.2 and 1.5 volts a.c. Make sure that no part of the voltmeter or your body touches ground when making this measurement! Once filament voltage has been confirmed, shut down the power supply, discharge the high-voltage feedthrough, and assemble the metal enclosure, making sure that it is completely sealed.

As a final check, connect the circuit to a heavy-duty 2- to 12-volt power supply (such as a battery charger) capable of handling 5 amperes. Connect an ammeter in the input lead. With a 2-volt input, the ammeter should indicate about 0.5 ampere. With 12 volts input, current should be about 2 amperes. The reading could go as high as 5 amperes if the circuit is loaded with a high-voltage experiment.

DANGER! There is at least 10,000 volts present on the top of the feedthrough insulator! Treat it with the greatest respect. Don't try to draw arcs with a pencil, and don't short this terminal to ground when the supply is energized. Also, don't touch the transistor case while the supply is operating.

Shut down the power supply, discharge the high-voltage feedthrough, and place the power supply in the cabinet as shown in Fig. 7, with the battery terminals and pilot light facing front.

Mark the point on the metal column that is directly opposite the top of the high-voltage feedthrough. Drill a ¼" hole at this point and deburr it. Obtain a length of high-voltage plastic tubing long enough to go from the top of the feedthrough to the center of the column. For still better insulation, insert another piece of 1/8" tubing inside the first one. Feed a length of #22 or smaller wire through this insulator leaving enough at one end to make a connection to the feedthrough and a small loop at the other (column) end, at the center of the crosshairs.

Attach a length of fine nichrome wire (obtained by dismantling an old wire-wound resistor) to the loop of wire at the bottom crosshair (wind the copper wire around the nichrome) and to the crosshair itself for support. Pass the nichrome wire up through the column and attach the top end to the upper nylon crosshair, making sure that the wire is reasonably straight and does not come near the sides of the metal column. Cut off any loose end.

(Nichrome wire is used here because the high voltage produces a tiny corona which would ruin copper wire but does not harm the nichrome. Steel wire can be used, but it will eventually rust and disintegrate.)

Do not use any mechanical device to connect the nichrome wire to the high-voltage lead since this joint will have to be disassembled occasionally so that the column can be removed for cleaning. Replace the plastic cover on the bottom of the column. Place the blower container on top of the column, making sure that the exhaust is toward the front of the cabinet. Secure this can in place by wrapping tape around the seam. Make a wire connection between the metal column and the metal chassis of the power supply by soldering at both ends.

The appearance of the stack can be improved by spraying it with paint, but don't get paint in the blower mechanism or on the high voltage leads.

Line Operation. The Transcipitor can be operated from a conventional low-voltage d.c. power supply such as that shown in Fig. 8. Mount the transformer on the small shelf in the cabinet and the filter capacitor on a clamp secured to the cabinet wall. A TV power socket is mounted on a small piece of metal and located on the cabinet wall so that power can be applied to the system only when the front panel is in place. A TV "cheater" supplies power to the socket and is mounted on the front panel aligned with its receptacle. The two rectifier diodes are mounted on a heat sink on the rear wall. Wire the power supply point-to-point as shown in Fig. 8.

If the fan motor is of the 12-volt variety, wire it to the power supply, observing proper polarity. If the motor is 117 volts a.c., connect it in parallel with the input to the power transformer.

Final Assembly. The finished project should now look like the one in Fig. 7. The column, with the fan at the top should just fit snugly within the cabinet. It may have to be wedged if too loose; or the cabinet top may have to be hollowed out slightly if the fit is too tight. Recheck all wiring, making sure that the metal column is connected to the positive battery input (at ground) on the supply. Make sure that the high-voltage feed is in the clear and that all parts attached to the cabinet walls are on tightly.

Obtain a length of flexible vacuum cleaner hose. The hose should be covered with plastic rather than cloth to make cleaning easier. Wash the hose thoroughly, inside and out, with a good detergent and then rinse thoroughly. Cut a hole in the front of the cabinet so that the end of the hose and the exhaust on the fan can be mated. It may be necessary to make up some type of size-matching device if the two are greatly different in size. Check all dimensions, and then mount the front panel on the cabinet using a few screws to secure it.

Operation. With the front panel in place and the a.c. supply connected, the blower should start up and moving air should be felt at the outlet. Hold a lighted cigarette or other source of smoke near the ring of holes at the bottom of the column. If everything is working properly, smoke will enter the column, but the air coming out will be clean with no trace of smoke.

For conducting delicate experiments or for drying paint on small items, another cabinet such as that shown in Fig. 9 can be constructed. The vacuum-cleaner hose is coupled to this cabinet; and a small vent in one wall permits the air to escape from the interior. The front door can be constructed with a glass insert and a light bulb can be installed within the cabinet for viewing experiments.

When it comes to removing pollen, dust, etc. from an area as large as a room, the Transcipitor will work—to an extent. It does not have the capacity to handle a very large room; but, in a small room, with windows and doors closed, its effect is quite noticeable.

Every so often, inspect the metal column for dust accumulation. Remove the column from the cabinet, hold it over a paper sack and remove the bottom plastic cover. Shake the column gently to remove dust particles stuck to the sides. Clean the inside walls before reinstalling the column in the cabinet. The stack can be inspected from the outside by shining a flashlight through the bottom array of holes and looking into the other holes. If you can see the dirt, empty the column.*

Example C

Parts List

C1—50-µF, 50-volt electrolytic capacitor

D1-D4—2-ampere, 50-volt diode (International Rectifier 20A05 or similar)

D5—1.5-ampere, 50-volt diode (1N4816 or similar)

F1—0.6-ampere fuse

I1—General Electric #GE 1819 28-volt lamp

Q1, Q2—MPS3702 transistor

R1—24,000-ohm

R2—20,000-ohm

R3—2000-ohm   All resistors ¼-watt

R4—27,000-ohm

R5—1000-ohm

R7—24 300-ohm, ½-watt resistor connected in parallel (see text)

R6—2000-ohm linear-taper potentiometer

---

*From Walde T. Boyd, "Build A Transcipitor," *Popular Electronics*, June, 1970. Reproduced with permission.

SCR1—C106F1 silicon controlled rectifier

T—2-ampere, 25.2-volt filament transformer (Allied Radio No. 54A4140)

TDR1—2000-ohm temperature-dependent resistor (Fenwall No. LP32J2)

1—5" x 4" x 3" metal utility box

Misc.—Control knob; a.c. line cord; rubber grommets; epoxy potting compound; hardware; hookup wire; solder; etc.

Construction. The heater element, $R7$, is a simple affair made up of twenty-four (24) 300-ohm resistors connected in parallel as shown in Fig. 1. To provide rigidity to the assembly, it is suggested that you "ladder" assemble the resistors between two heavy-duty wire busses.

Although the heater arrangement is rated at only 12 watts in free air, it will safely dissipate 50 watts of "heating" power when submerged in water.

Since the heater element is to be operated completely submerged, it must be water-tight. So, after assembling the element, carefully check the heavy wires you plan to use between it and the control/power circuitry for nicks and holes in the insulation. When you are satisfied the wire is safe to use, solder a 5'-10' length to each of the heater element busses.

Now, coat the entire assembly and 2" or 3" of the wire with epoxy potting compound. (Use only a true epoxy, one that must be prepared from separate resin and hardener compounds immediately prior to use.) Do not make the coating too thick, but make certain that the entire assembly and the attached ends of the wires are completely sealed. A water leak from improper sealing will cause the heater to fail, and copper in solution from the wires will harm your fish.

After the first application of epoxy has set (wait at least 48 hours), put on a second coat and wait for it to set. If the outer coat is not completely set, it will allow volatile solvents to enter the aquarium water—obviously also harmful to your fish.

The temperature sensor, $TDR1$, is also operated while submerged in water. Consequently, the same steps must be taken in selecting interconnecting wires and epoxy potting it as above. When both assemblies are finished, they should appear as shown in Fig. 2.

The layout of the power supply/control circuit (see Fig. 3) components is not critical, permitting any type of chassis wiring you prefer. For your convenience, an actual-size printed circuit board foil pattern and component layout guide are provided in Fig. 4.

When mounting transistors $Q1$ and $Q2$, locate them close together, but not touching, to minimize thermal differences in their base-to-emitter junctions. A small heat sink might be needed for *SCR1*; hence, its tab is shown bolted to the angle bracket. (If you substitute another type of SCR for the one specified in the Parts List, check its specifications to make sure that less than 500 micro-amperes at the gate will drive it into conduction.)

When all components are mounted on the circuit board, mount the board, transformer, fuse holder, potentiometer, and pilot lamp inside the utility box as shown in Fig. 5. The center-tap lead of the transformer can be cut short and the stub taped.

Twist the sensor and heater element wires together and route them and the line cord through rubber-grommet-lined holes in the rear of the utility box. Tie strain relief knots in both cables inside the box, and interconnect all components and assemblies. Assemble the box.

Calibration and Use. Immerse the heater element sensor in a glass of cool water. NEVER operate the system unless the heater is immersed in water, preferably with the sensor in the same water. Plug in the line cord; the pilot lamp should immediately come on, indicating that the system is operating. In a few minutes, when the water heats up, the light should extinguish. Rotating the control knob clockwise should cause the light to come on again, counterclockwise to extinguish it. If the reverse happens, unplug the line cord and reverse the connections to the outer lugs of the potentiometer.

A thermometer of known accuracy is needed to properly calibrate the system. First immerse the sensor and heater in about a pint of cold water. Set the control fully counterclockwise, and plug in the line cord. Now stir the water constantly with the thermometer. As soon as the lamp extinguishes, remove the thermometer from the water and note the temperature indicated. Record your reading on the front of the utility box, in line with the index of the control knob.

Return the thermometer to the water and advance the control until the lamp just comes on again. Stir the water with the thermometer until the light again extinguishes. Record your reading. Continue this process until you have enough calibration marks. Then disconnect power from the system, and use a decal or dry-transfer lettering kit to finish the front panel.

In use, the heater element should be buried just under the surface of the gravel and/or sand in the bottom of your aquarium, in a location where the circulator can feed the water over it. Leave the sensor suspended in the water 2" or 3" "upstream" of the heater element. If desired, the sensor element can be camouflaged by the tank plants. Then plug in the line cord and set the temperature control.

The electronic aquarium heater has more than sufficient power for the standard 15-gallon aquarium. It will also serve a much larger aquarium if the water temperature is not to be too much greater than the ambient room temperature.*

Example D: Results Section of a Technical Report

Tests Performed

To determine the effectiveness of the ferroresonant transformer in protecting color-TV receivers, an independent testing laboratory ran a series of tests.

The results clearly indicated that the TV set with Colorvolt was affected the least when the various appliances were initially started up. The worst case example was when a ten-inch table saw was turned on which caused the voltage input to the set with the Colorvolt to drop from 117 volts to between 104 and 107 volts and the unregulated set to drop to 89 volts. Under low voltage line conditions (106 volts), the effect was more severe, with the input to the set with the Colorvolt only dropping to 99.6 volts and the unregulated set to a low 76 volts.

The results of additional input/output tests conducted to determine the voltage regulating capability of the Colorvolt are listed in Table 1.

Assuming then that the radiation problem is directly proportional to input voltage going to the high-voltage section of the receiver, Colorvolt would alleviate the radiation threat by pulling voltages as high as 130 volts a.c. down to within the nominal performance range of most new color sets.**

CONCLUSION

The final or terminal section of the technical report is analytical and interpretive. It discusses, concludes, and recommends on the basis of results obtained with the materials and methods reported in the body of the report. The technical reporter becomes commentator, discussant, recommender, and concluder in the end section and tries to offer a final, broad interpretation of the specific result or outcome recorded in the body section.

---

*From Stacey Jarvis, "Electronic Aquarium Heater," *Popular Electronics*, January, 1970. Reproduced with permission.

**From Neil Ferency, "Ferreresonant Transformer Improves Color TV," *Electronics World*, February, 1970. Reproduced with permission.

Since the body of the report attempts to restrict itself to objective data pertaining to materials, methods, and results of the study, the writer is not free to comment about the various meanings which may apply. In the conclusion section the opportunity exists for exploring briefly what the results seem to mean and how they may be applied. On the basis of such discussion and commentary the writer is able not only to draw conclusions but to make recommendations for action to his readership. One naturally expects that all discussion, commentary, recommendations, and conclusions will be related logically to the introductory and body sections of the report.

Materials presented in the different subsections of the concluding part of the report are really what the report is all about. The report, of course, has no value in itself—the report is obviously a means to an end, that end being informed knowledge as the basis for action. No writer should ever lose sight of the idea—a report is not a beautiful thing in itself. It is a useful, necessary, perhaps invaluable instrument for seeing a certain set of facts in logical, coherent, orderly, and clear fashion. But it is the uses which the intelligence may make of these facts which are basically important. By analytically weighing the possibilities, thoughtfully drawing conclusions, and carefully developing recommendations, the technician-writer is providing his greatest service to the reader.

Example A: Discussion Section of a Technical Report

Advantages and Disadvantages

There are many presently popular approaches to decimal counting, decoding, and readout. Irrespective of a manufacturer's claims, no one counting system is "ideal" nor is any one system suitable for all possible counter and display applications. In the case of this system, there are both advantages and disadvantages to its use.

Its greatest advantages are its simplicity and the add-subtract feature. Only two or three parts—an IC, a readout, and possibly a socket are needed per decade. Assembly on a single-sided, multiple-decade PC board with only 13 holes and two jumpers per stage is possible, especially if a slightly wider-than-normal numeral spacing is acceptable.

The Add-Subtract feature is offered on very few competing systems, and then only at considerably higher cost and complexity. On the other hand, the subtract operation is rarely used in digital instruments and is accomplished in a different (parallel) manner in calculators and computers; it is unessential for practically everything but predetermining counters, positional controls, and very simple arithmetic operations.

The display is a pleasing green and is visible over a wide angle. It does not have exceptional brightness and a totally dark area behind the display is recommended, along with a filter. The "boxiness" of the characters is probably objectionable to a "Nixie"-oriented instrument market. A more legitimate objection is the off-center "1" and awkward "4" presented by the display, and the resultant "holes" in a numeric sequence. This may be overcome by using a nine-bar display and an external transistor or two. The output voltages and currents of the IC are only compatible with vacuum fluorescent readouts.

The greatest disadvantages are undoubtedly the rather "weird" supply voltages required: −27 volts, −13 volts, and a 1.6-volt, 45-90-mA filament voltage *referenced* to the −27 volts. Simple translators have been included in the assembly. Each of these consists of a *p-n-p* transistor and two resistors and allows the assembly to be driven from conventional 0- to +3-volt RTL or DTL logic signals.

Other disadvantages include the limited speed which is partially offset by today's low-cost decade scaler circuits. The slower speed does buy better noise performance, particularly in industrial environments. The IC, being a MOS type, can be damaged instantly by reversed supply power, extremely careless handling, or by very large line transients. Protection has been included both inside the IC and on the prototype assembly for normal static, installation, and handling.

The readout has several failure modes, and thus has a limited, but probably quite acceptable life. At least one readout manufacturer has "beefed up" its readouts considerably to make them more immune to vibration. Sockets are probably a worthwhile addition for heavy-use applications.

The readout is also electrostatic-sensitive, with the glow dancing around if you bring a finger near the display. A physical barrier (such as a color filter) and some anti-static spray takes care of this particular problem. A newer *Tung-Sol* readout is available that is smaller, more rugged, and not electrostatically sensitive.

One big objection experimenters and technicians will have is the steep pricing structure of both the IC and the readout. Only in very large quantities can the counter be built for less than $7 per decade. In small quantities (1-99) cost will run around $29 per decade ($24 per IC and

$5 per readout). Commercially available competing decimal-counter kits presently range from $10 to $30 per decade in small quantities, depending upon the readout and the speed of operation; none of these is as simple or has the subtract feature.

A block diagram of the MEM1056 is shown in Fig. 1. This is a MOS integrated circuit and comes in a 24-lead flat pack that operates off two negative supply voltages of −27 and −13 volts. The input logic swings from 0 volt ("0" or "No") to −27 volts ("1" or "Yes").

There are four main parts to the IC: the up/down decade counter, the storage register, the decimal-to-seven-bar logic converter, and the output drivers. The up/down decimal counter can accept either *series* (a sequential series of count pulses) or *parallel* (simultaneous appearance of count pulses) data.

To count serially, 0- to −27-volt pulses are applied to the Clock input. The clock is a "two-phase" type. The number in the counter changes as the Clock input goes from −27 to ground. Some first-decade conditioning is needed; the Clock input signal must have rise and fall times faster than 20 microseconds, and all mechanical contact or push-button inputs must be made "bounceless" to prevent contact noise from causing erratic operation.

The serial-input clock pulses are controlled by the Count Disable and the Count Down inputs. If the Count Disable input is at −27 volts (logic "1"), all input clock pulses will be ignored. If the Count Disable input is grounded (logic "0") input pulses are accepted. The Count Disable input is used to gate the counter for frequency or period measurement. The readout that is used remains lighted constantly irrespective of the condition of the Count Disable input.

The Count Down input determines whether the input clock pulses will be added or subtracted from the tally inside the IC; −27 volts ("1") on this input makes the IC subtract and grounding ("0") this input makes the IC add.

The Preset input forces the counter to the 0 count when −27 volts ("1") is applied, and does nothing when grounded. The "1," "2," "4," and "8" inputs may be used in combination with the Preset input to enter a number in parallel. These inputs are normally left grounded. To enter a "7," the Preset and only the "1," "2," and "4" inputs are lowered to −27 volts ("1"). The proper combination (a BCD word) is used for each required count. In this manner, a keyboard or a selector switch can enter counts without generating a separate series of pulses for each count. The readout will automatically produce an "F" indication if a false count (binary 10 through 15) is entered.*

Example B: Recommendation Section of a Technical Report

Self-Cycling Circuit

Both of the circuits discussed above suffer from one disadvantage—the push-button must be depressed and released for each cycle of sound. An obvious improvement would be automatic cycling and one way to do this would be to replace the push-button by a timer-controlled latching relay. Another method would be to use an astable multivibrator or flip-flop to generate a long pulse of voltage, followed by a period of zero output, that can be applied to the input in place of the switch.

However, the approach used here was to eliminate the time-delay capacitor and resistor and apply the output of a triangular-wave generator (Fig. 2A) to the input of the oscillator circuit. The rise and fall of voltage at the output of the triangular-wave generator provides the proper bias current for the siren sound.

The triangular wave is derived from a unijunction relaxation oscillator that generates a sawtooth voltage waveform. The UJT relaxation oscillator ($Q1$) acts like a neon-bulb oscillator and does not conduct until the voltage on its emitter reaches a critical (peak-point) value. The length of time required for this to occur is dependent on the time needed to charge capacitor $C2$ through resistor $R1$. When the proper voltage is reached, $Q1$ conducts, discharging $C2$; and when the voltage across $C2$ drops below the critical value, $Q1$ stops conducting, starting the cycle over again. This repeated cycling produces a sawtooth voltage at the emitter of $Q1$.

To adapt the sawtooth waveform generated by the UJT to the slow rise and slow fall needed to produce an authentic siren sound, diode $D1$ and capacitor $C3$ (Fig. 2A) are added to the relaxation oscillator. The diode is forward-biased (conducting) during the period when $C2$ is charging so that $C3$ charges at the same slow rate as $C2$. It is this slowly rising triangular waveform, fed to the oscillator circuit (Fig. 2B), that causes a rising tone to be generated by $Q2$ and $Q3$. However, when $Q1$

---

*From Donald E. Lancaster, "Add–Subtract MOS IC Decimal Counter," *Electronics World*, June, 1970. Reproduced with permission.

(Fig. 2A) conducts and $C2$ discharges, $D1$ becomes reverse-biased (stops conducting), causing $C3$ to discharge slowly through $R2$ and $R3$. This action causes a falling tone, generated by $Q2$ and $Q3$, which continues to fall until the rising voltage on $C2$ is greater than the voltage on $C3$ (actually about 0.7 volt greater to overcome the threshold of the silicon diode), when the cycle repeats. The rise time of the triangular waveform is controlled by $C2$ and $R1$ while the fall time is controlled by $C3$ and $R2$.

Obviously, there is no need for a push-button when using the triangular-wave generator discussed above since the siren cycle will continue to provide a loud sound until power is interrupted.

The triangular-wave generator shown in Fig. 2A is for 12-volt, high-power operation, but can also be used with the lower power oscillator of Fig. 1A. As shown in the lead photo and the photo below, the entire panic-button circuit can easily fit into a small chassis box.*

Example C: Conclusion of a Technical Report

Where different parts are packaged in a single container, gated automatic feed of parts may be controlled by a counter and associated programmed output circuits. Multi-position machining operations may be controlled by friction drive wheels at different locations along a length of stock. The wheels deliver pulses to the counter, which is programmed to provide signals at the desired locations. Using two counters, one for the x-axis and one for the y-axis, the machining operations can be located anywhere in the x-y plane.

In multi-valve batching, several valves control the influx of substances into a mixing region. A rotating volume meter delivers pulses to the counter, which indicates the quantity of flow, and actuates the valves. A timed sequence can be generated by using an oscillator to deliver pulses at a specified rate to the counter. The resulting signals (of various interval lengths) are directed to the appropriate devices.

A high-speed predetermined counter has been used in a packaging operation, requiring rates to 200 Hz. A set of pneumatic position switches selected the number of pieces per package. The counter controlled the entire packaging operation.**

---

*From Paul Franson, "Panic Button," *Electronics World*, May, 1970. Reproduced with permission.
**From Robert F. O'Keefe, "Fluidic Decimal Counter," *I & CS*, June, 1969.

VARIATIONS IN ORDER OF PARTS

Writers of reports today have responded intelligently to needs of readers by adopting streamlined modes of ordering the different parts. Current practices recognize that different readers of a report have separate reasons for consulting it. Not all report readers are trained to, or interested in, the detailed technical aspects. To require all readers, therefore, to struggle through the full introductory and body sections before reaching the recommendations and conclusions which are chiefly important to them is both useless and self-defeating. Several different orders of presentation, accordingly, are increasingly common in technical reports encountered in schools, business, and industry.

Perhaps the most familiar device for serving the reader better is the abstract that appears as the first item of business in many reports. The abstract serves to summarize concisely the general scope of the report and to indicate to the reader whether reading of the full contents is necessary. Reports utilizing an abstract usually proceed beyond that point to the conventional order of introduction, body, and conclusion. But inclusion of the abstract does not preclude departure from the traditional order of parts.

Some reports containing abstracts, for example, present immediately after the abstract a synopsis of the general conclusions and recommendations, rather than withhold these items until the end. From that point, the report may proceed to introductory background information and then on to the body containing materials, methods, and results. More detailed presentation of recommendations and conclusions may then appear as the terminal element.

There are reports, either with or without an initial abstract, that present results before anything else. This practice would be particularly appropriate in an experimental study the results of which appear to be of chief interest to the potential average reader. A report so ordered might then proceed to materials and methods before considering the discussion, recommendations, and conclusion. Or it might with equal logic move from results to recommendations and conclusions before taking up detailed aspects of materials and methods.

A most useful element which appears in numerous technical reports is the summary or recapitulation, as the final section. The writer aids the reader by briefly and clearly presenting a thumbnail sketch of the highpoints of the full report. In this way the reader can gain a final overall view of the picture and make an integrated synthesis of the knowledge.

All of these variations in the order of the parts are suitable reminders that reports are written by people for people. Though both logic and practice indicate the general need for beginning, middle, and end in a report, blind unthinking adherence to a mechanical order of presentation is undesirable in the extreme. The alert technical writer is always aware of the special nature of his particular report in an individual situation. He continues to explore critically in his own mind the various possibilities available to him in ordering his parts for maximum clarity and effectiveness. Chiefly, he strives to adapt his report to what he assumes are the major needs and interests of his readers. Personal inventiveness and imagination are not enemies of technical writing.

ELEMENTS OF THE REPORT IN REVIEW

Considerable paraphernalia and elements may accompany or be part of a long technical report. The following listing is by no means comprehensive. It aims to include merely those most basic elements likely to be part of a simple report prepared for ordinary use internally or externally.

Cover. Since any report may have reference value, the cover is needed to identify and protect it. Author, title, completion date, and identifying numbers, series, and departments appear on the cover. Some organizations provide printed forms for information to appear on the cover which catalogs and identifies the report according to an established company system.

PARTS OF THE REPORT

Title page. A title page contains the title, subtitle, author with his professional title and location, name and location of the unit or division for whom the report is prepared, date of completion, and serial designation, contract, or project numbers where appropriate.

Letter of transmittal. The letter of transmittal is addressed to the authorizer or requester of the report from the writer. The letter may accompany the report after the title page, or it may be sent separately. Such a letter usually recapitulates the facts surrounding the mechanics of authorization of the report so that these may become part of the record. The writer may choose to review in this letter the scope and history of the project, to suggest the range and limitations of the report, to indicate its outline and possible uses, or to survey briefly some of its conclusions and recommendations.

Table of contents. The table of contents is always arranged and presented in the form that will most effectively provide the reader with a total view of the individual parts of the report in relation to one another. For this reason, the table may be indented, presented with headings or subheadings, or even be presented in outline form. Numbers referring to specific pages on which indicated sections commence are usually listed in a report of any length as an aid to the reader. The table of contents will indicate where the list of references begins, and will do likewise for the appendix. Following this will be a list of tables and/or a list of figures. All efforts are made to offer a table of contents as helpful and useful to the reader as possible.

Synopsis. Following the table of contents as the first element of the body of the report proper will be the synopsis, if one is to be included. The synopsis is variously referred to as an abstract, precis, resume, digest, or summary, and the student may consult a standard dictionary to determine the different shades of meaning attaching to each of these terms. In general, the synopsis attempts to present a brief general view of the major ideas and facts contained in the report as these might serve most effectively the needs of a particular reading audience.

Appendix. The appendix follows the body of the report and contains whatever detailed information the writer judges necessary to support and verify the accuracy and authenticity of material in his report. The tendency today is to remove most tables and graphs of any length from the body of the report to the appendix, so as not to interrupt the flow of reading but to allow opportunity for detailed study later. An appendix may also include copies of relevant correspondence relating to the report. Each item in the appendix is labeled to correspond to the listing shown in the table of contents.

Bibliography. A bibliography is a list of all the printed works used in the preparation of the report. It is offered as an aid to the reader and a means to checking the accuracy of cited information. In many technical reports, extensive reference to, and citation from, the technical and scientific literature constitutes an important part of the writing.

a supplementary view...

SUGGESTIONS FOR THE PREPARATION OF TECHNICAL PAPERS *By Robert T. Hamlett*

From Proceedings of the IRE, *March 1950. Reprinted by permission of the Institute of Radio Engineers and the author.*

I. INTRODUCTION

The engineer rarely faces a more clear-cut opportunity for accomplishment than that presented to him when he is chosen to prepare a technical paper. The direct benefits of successful accomplishment are threefold: the author's professional prestige is enhanced, the reputation of the organization he represents is maintained or improved, and last but by no means unimportant, the standing of the engineering profession in general is raised. With these inviting benefits, it is unfortunate that they are only occasionally realized because of poor preparation and even poorer presentation of the technical paper.

The engineer has always labored under the stigma that "Engineers cannot write." It is questionable whether engineers as a class write any more poorly than doctors or lawyers or salesmen. Perhaps the subjects we write about are more complex and require more specific knowledge of the reader. Whether this poor reputation is justified or not, the only logical course for engineers is one of continual self-improvement until this undesirable *class distinction* disappears.

Courses in technical writing are given in many of the better schools but unfortunately the student seldom appreciates at that time the importance of effective writing, and even worse he retains little of what he learned because in the years immediately following graduation there are few opportunities for him to prepare a technical paper. His knowledge of ordinary rules of grammar, rhetoric, and logical presentation become rusty from inactivity, and he finds that writing clearly and keeping in mind these rules is like reading a foreign language taken in high school; the rules tend to confuse rather than simplify his task.

To attempt to lay down in this article a complete and final set of rules for preparation of the perfect technical paper would be an impossible task. There are many variables entering into the preparation of a particular technical paper to be presented under certain circumstances at a specific meeting. Further, the complete skills involved in preparation of a paper encompass the entire education and experience of the engineer. However, there are certain accepted qualities which any successful technical paper must possess. It is the purpose of this article to refresh the engineer's mind on some of these fundamentals and to stress other factors which can make his paper more effective.

The material for this article is derived from the author's avid interest and attendance at technical meetings, from the instruction pamphlets of prominent technical societies, from a number of excellent textbooks on technical writing . . ., and from the helpful suggestions of fellow engineers.

II. THE PRINCIPAL ELEMENTS

A. THE OUTLINE

It is well to recognize at the beginning that writing a technical paper is hard, and sometimes very boring, work. There is certainly no royal road to perfect technical exposition. One must be willing to write and rewrite many times. The successful writer often tears up his copy and starts over again when he finds that the logical development of the paper is blocked by the existing approach.

Most authorities agree that the best way to start is by setting down an outline of the paper, i.e., writing down the principal topics to be covered. Carry the outline as far as possible the first time, let it rest for a few days and then try again. Missing blocks in the outline will begin to appear with increasing ease. The major and minor topics will form a basis for the start of actual writing. Do not worry about organization of the paper until a large portion of the text matter has been written. Preparing an outline, a first draft, and a final copy may appear to involve an unnecessary amount of work, but it is usually true that such a routine actually saves time.

When the basic technical material has been developed it is time to look at the paper from the reader's viewpoint. The reader's requirements are simple but definite: he must be carefully introduced to the subject of the paper, the subject must be adequately covered, and finally the subject must be concluded. It is alarming how often technical papers violate these three simple rules.

The development of a good technical paper may be compared to the preparation of a good dinner. First there must be an appetizer (introduction) which whets the reader's interest in what is to follow, second the main course (body of text) must be well balanced and full of meat, and third the dessert (conclusion) must be satisfying and should leave a pleasant effect on the reader. While many other courses (soup, salad or spinach!) may be added to round out the technical meal, these three basic elements remain the same for any paper, and must be blended together carefully to accomplish the writer's purpose.

I am indebted to one of my Navy Publication friends for an apt phrase in this connection. He says every effective piece of technical writing requires "that first you tell them what you are going to tell them; then you tell them; and then you tell them what you have told them"; this simplified expression repeats again the basic requirement in any technical paper for *introduction, main body* of information, and *conclusion.*

B. THE INTRODUCTION

Without question the *introduction* is the most important part of the paper—from the reader's viewpoint. Whether the reader will continue with the paper at all depends largely upon the impression created by the *introduction.* Because of the tremendous growth in variety and complexity of technical subjects, there is an increasing demand from readers that the first page or two of a technical paper should provide a comprehensive idea of the whole paper. The average writer is likely to write too long an introduction or none at all.

It should be recognized that while the *introduction* is read first, it should be written last—after the main body and conclusion are completed, for it must include in an abbreviated form some of the material from each. Do not hesitate to spend a large amount of time in the preparation of the *introduction* for it will pay attractive dividends in number of readers.

C. THE MAIN BODY OF TEXT

This portion of the paper contains the technical facts which justify the paper itself. This part of the paper offers the least difficulty to the engineer. He is on more familiar ground where technical grasp of the subject is the primary requisite. If an outline has been prepared, the writing should proceed satisfactorily. The first rough draft should be written rapidly without regard for literary style. Too much attention at this time to grammar and spelling will slow down the development of basic materials.

A search of contemporary literature on the subject should be made so that the material to be presented will not unknowingly duplicate or contradict existing literature. If the paper does differ in important conclusions with any previously accepted literature, the differences should be pointed out and substantiated by the author. The author should make use of the facilities offered by his engineering library for a search of contemporary literature on his subject. The preparation of a satisfactory bibliography is covered in another portion of this article.

Accuracy of data in the paper hardly needs mentioning. The engineer by nature and training is careful in the weighing and analyzing of data and is seldom tempted to distort facts to gain a temporary advantage. However, he cannot exercise too much care in being correct and honest in all of his statements.

Be constructive and positive in presenting the material, never antagonistic, pessimistic, or negative. Tearing down some other engineer's reputation will seldom add to the author's professional standing. Direct criticism of competitor companies by name is particularly unwise. In fact the shortest route to the listener's good graces is by paying tribute to others, whether they are competitors or associates.

While it is essential that the text cover the subject adequately, it is also important that it be neither too detailed nor too complex for the intended reader. After the main body is prepared, go over it several times to cut out material not absolutely necessary for clarity. Almost any technical paper can be boiled down considerably with little loss to the reader. It is an old story around Sperry that our former president, Mr. Reginald Gillmor, was a stickler not only for good written material but also for concise writing. Many times he would return copy to the writer with a notation "cut it in half." After sweating it out the writer would make the required reduction, but then get another shock when he received a second note from Mr. Gillmor "to cut it in half" again. While this method cannot be applied generally, many technical papers could be cut in half and be more interesting and just as informative.

The writing in technical papers should be impersonal; do not use *I* if it can be avoided; try to keep the language in the third person. It is permissible, however, to use *we* occasionally, if its meaning is clear. For example, following several references to a project in the author's company, it may be more diplomatic to use *we* instead of repeating the company name and be criticized for too much "name advertising."

Sentence length is important in the technical paper. When the draft copy has been completed, it is advisable to go over the sentences again and separate the longer ones into lengths that will not burden the reader's power of concentration. . . .

Carelessness in spelling, grammar, or speaking by the engineer may bespeak carelessness in other elements of the paper and may well lead the audience to question the accuracy of the technical statements. Do not split infinitives when you can avoid doing so. The prejudice against split infinitives is deep-seated and persistent. Usually it is just as easy to write effectively as it is to effectively write. However, if there is real gain in emphasis or clearness through splitting the infinitive, you can do so and be in the company of many excellent writers—but you are likely to be misjudged by some readers.

The use of headings and subheadings is often neglected by the technical author. The more complex a subject becomes, the more necessary it is to break it up into a number of parts which the reader can visualize. More than three degrees of subheadings are not recommended for a paper. For instance, a good example of main and subheadings would be:

I. General description of xx radar set
   A. Transmitting system
      1. Purpose and general description
      2. Detailed circuit analysis
         a. Modulator
         b. Pulse forming network
         c. Clipper circuit
         d. _____
         e. _____
   B. Receiving system
      1. Purpose and general description
      2. Detailed circuit analysis
         a. Local oscillator
         b. Receiver mixer circuit
         c. I-F section
         d. _____
         e. _____

Avoid the use of unfamiliar terms unless you have time to define them. . . . If necessary, a list of symbols should be provided to clarify the text. Long equations or complicated derivations should be placed in an appendix, rather than in the main body of the text. Use footnotes sparingly; from the reader's standpoint it is much better to integrate such material with the text. Bibliographical references are an exception; they are usually carried as footnotes on the same page of the matter to which they apply.

   D. THE CONCLUSION

The *conclusion* is another challenge to the writing ability of the engineer. It should sum up the major points made in the text and leave the reader with a feeling that the conclusions are fully justified by the data presented. The normal purpose of the technical paper is to inform and not to sell or arouse to action, but it is difficult to visualize a good paper which does not accomplish these latter purposes to some degree. The *conclusion*, like the *introduction*, requires careful writing and rewriting before it will accomplish the author's purpose. One simple warning: do not state that certain conclusions are "obvious"—nothing irritates the average reader more than the assumption by the writer that his own logic requires no substantiation.

E. ILLUSTRATIONS...

Illustrations can add much to the readability and conciseness of technical papers. As Lord Kelvin pointed out "a single curve, drawn in the manner of the curve of prices of cotton, describes all that the ear can possibly hear as the result of the most complicated musical performance...." The judicious use of illustrations will improve the paper in many ways. It is well to review the completed text for instances where illustrations can shorten or supplement the written material. Do not try to show too much on *one* illustration; a simple illustration will be instantly valuable to the average reader who may not be inclined to concentrate on more complicated diagrams.

F. BIBLIOGRAPHY

Every good technical paper should have a bibliography of literature related to the subject. This is necessary not only to guide the reader if he has desire to pursue the subject further, but it also indicates that the author is acquainted with the literature in his field and has made use of others' knowledge in the preparation of his paper.

Engineers in preparing their papers frequently and inadvertently offend their readers by using incomplete bibliographical references. In the case of books, the reader may wish to procure for his personal library one or more of those listed. It is appropriate then to include the publisher's name. Page references also are valuable, and page references usually are erroneous unless the edition number of the book also is given. In the case of periodicals, it is helpful to list the volume number as well as the month and year along with page references. Libraries bind their periodicals into volumes and it is helpful both to the reader and the librarian when this number is known. Bibliographies are usually carried as footnotes on appropriate pages of the prepared copy but may be included as a separate section after the *conclusion* section.

The forms of bibliographical references may vary but the following are typical and adequate:

For a book:

1. J. H. Morecroft, "Principles of Radio Communication," John Wiley and Sons, Inc., New York, N. Y., 3rd Edition, p. 402; 1933.

For a periodical:

1. P. H. Trickey, "Field harmonics in induction motors," *Elec. Eng.*, vol. 50, pp. 937–939; December, 1931.

IV. CONCLUSION

The preparation of a good technical paper is a real challenge to the engineer. Into its preparation can go the complete range of his abilities—education, experience, and knowledge of human behavior. The technical paper sticks out all over with its good and bad points. No amount of patience and concentration is too great to apply to the task, and the rewards always justify the effort.

The accompanying check list may serve as a "silent" critic of a technical paper.

Good organization, accurate and complete technical material, correct grammar and spelling, suitable illustrations, and effective delivery—these basic points should be kept in mind as the principal factors which will make the technical paper command the interest of its audience, which after all is the only justification for writing it.

Check List
For the Preparation . . . of a Technical Paper

Do—

Recognize the personal and professional opportunities presented in the preparation of a good technical paper.
Prepare an outline before beginning actual writing.
Be willing to write and rewrite every part of the paper.
Be extremely careful with the accuracy of your material.
Consider reader's viewpoint carefully.
Be sure the paper has clearly defined *introduction, main body,* and *conclusion,*
Write the *main body* first, the *conclusion* second, and the *introduction* last.
Keep the main text as concise as possible.
Put long equations and derivations in an appendix.
Use headings and subheadings for complex material.

Prepare a *conclusion* that sums up the main points made in the body of the text.

Use adequate and suitable illustrations. . . .

Give proper credit to any individuals who inspired or contributed substantially to the paper.

Don't—

Use first person; third person is preferable.

Make mistakes in spelling or grammar.

Split infinitives—unless you are sure it helps!

Employ long and complicated sentences or paragraphs.

Use unfamiliar symbols—if they must be included, define them.

Include too many footnotes; integrate them with the text.

Assume your conclusions are obvious to the reader.

Hesitate to write and rewrite the paper several times.

Use illustrations that have too much in them. . . .

Chapter 9
PRESENTING THE COMPLETED REPORT

All of the work in planning, researching, writing, and developing the report moves toward completion and presentation of the final copy to the reader. In the last analysis, success depends upon what goes into the final copy and how the format is presented. While one would think it unnecessary to say so, care and time are needed for the preparation of the completed manuscript. Beginning writers in particular sometimes defeat all of their otherwise admirable work by presenting to the reader a poorly typed, incorrectly set up report. Whether the technician does his own typing or not, it is he who must bear full responsibility for errors in spelling, poor grammar, or other faults. Since so much depends, therefore, upon this last stage of the work, every effort should be made to make the manuscript perfect in every detail of presentation.

FINAL DRAFT

It is foolish even to think that a final copy for the reader can be prepared directly from the writer's research and bibliographical notes. A working final draft is needed. While the completed report itself will be typewritten, some technicians will prefer to make their final draft in longhand on legal-size paper. If the draft is in typescript, it is best to triple space between lines to allow for the many revisions, insertions, and alterations which will surely be necessary.

Since the final plan of the work will probably have changed from that of the preliminary outline, a final working outline should be prepared at once. In researching his material and working his notes, the writer will frequently discover new information and better ways of arranging his information. While some preliminary outlines will hold through to the end, the writer should in no way feel bound to his earlier projected plan—it is merely a point of departure. A fresh outline, incorporating all of the newly acquired outlooks and insights to the subject of the report, is a good start on the final draft.

Thereafter, with all of his note and bibliography cards, and with his outline to guide him, the writer does his draft. Transitional paragraphs or sentences to join the separate pieces of information on each card will be written as he goes. Numbers will be provided for the citations, and the footnotes will be roughed in in correct form at the bottom of each working page. A final bibliography will then be laid out in correct form.

Following this, a time must be set aside for careful revision. Every detail of punctuation and spelling and grammar must be verified. All of the factual information must be checked, including the correctness of the form of the footnotes and bibliography. The headings and subheadings must be clearly marked for final type style and/or indentation which will be given to them in the final copy. Then, graphic materials must be decided upon and prepared. References to the visualizations must be incorporated into the text at the appropriate points. A table of contents will be organized, and a trial cover page inserted. Only after all of this is the technician ready to type or have typed the completed report for presentation.

FINAL COPY

There is no prevailing manual of style which may be applied to the format presentation of all reports. Individual schools, companies, and organizations will have their own practices to guide the new student or employee in the preparation of written materials. Since a report is a stand-in for the man who prepares it—that is, it stands for him in the eyes of anybody reading it—a good impression can be gained for the writer who knows and follows prevailing practices. Overall appearance of the manuscript is a proper reflection of the writer.

Good quality and good weight unlined white bond paper of standard 8½ by 11 inches size is the general rule. Only one side of the page is used. Double-spacing is the general practice. Generous margins are provided at the top and bottom and sides of the page. Plenty of white space on a page is comforting to the reader and makes a good visual impression if the elements on the page are neatly indented, headed, and arranged.

If more than four or five copies of the report are needed (usually the case), a stencil or plate will be prepared for duplicating purposes. It is increasingly common today to prepare a single clean copy which is then photographically reproduced in quantity by one of the many excellent machines available. If this is to be the case, a fresh ribbon in the typewriter and clean type on the machine help assure better copies. Since colors are not reproduced in copying, graphics must be prepared in black and white only for clearest duplication.

All pages are numbered but the title page. Any preliminary pages—accompanying letters of authorization, approval, acceptance, etc.—are numbered in lowercase roman numerals (ii, iii, etc.). These numbers are centered about one-half inch from the bottom of the page. One begins numbering with ii, since the title page is i although no number appears on it. Page numbering for the rest of the report through to the end is with unpunctuated arabic numbers. Numbers of first pages of opening sections, if it is a long report, are centered about one-half inch from the bottom of the page. All other numbering is placed about one-half inch from the top of the page in line with the right margin. If one wished to be absolutely correct and formal, a blank page unnumbered could be inserted between the cover and title page and between the last page and the back cover.

In a long report main divisions may be referred to as sections, chapters, topics, or parts. Consistency is necessary, whichever term is adopted. It is customary to start each main division with a new page. Numbering and heading, of course, will conform to the listing in the table of contents.

QUOTATION

Material quoted directly from a printed source may appear in any technical report. There is no principle to follow in determining how much directly quoted information should appear in a report, but any writer would naturally wish to keep it to the absolute minimum. As was pointed out earlier, footnotes must be used for any ideas or information borrowed from an outside source, but only if the words are somebody else's are quotation marks required.

Especially in science and technology, where precision is of urgent importance, quotations must correspond exactly to the original source in spelling, phrasing, punctuation, and all matters. Inaccurate use of another writer's words is a serious failing. If the technologist has carefully recorded information on his research note cards in the manner described in an earlier chapter, there should be no great problem here. The first word in a short quotation need not be capitalized, even though it was so in the original, if the report writer links it grammatically with his own text. For example: The Smith study found that "choice of the proper solvent determines the durability of the finish." Capitalization of the original is retained in the quotation if the material is introduced by a complete sentence: "Choice of the proper solvent determines the durability of the finish."

A comma or period always appears inside the final quotation marks, whether or not a comma or period was part of the original source. This is the standard way for signaling the end of a quotation along with the quotation mark. Semicolons and colons used in conjunction with quotation marks are always placed outside the closing marks at the end. If they are part of the quoted matter, exclamation points and question marks are placed inside the final marks; they are otherwise placed outside.

It has been pointed out that double-spacing is the rule for typing the final manuscript. However, a quotation that exceeds four or five lines of type is single-spaced and indented from the rest of the text on the page. Since the guide words of the text and the spacing and indentation show the reader that it is quoted, quotation marks are not needed. The footnote number appears at the end of the quotation. Shorter quotations are incorporated into the body of the regular text, however, in the manner outlined in preceding paragraphs.

To reduce the amount of quoted material, it is permissible to omit words, phrases, or sentences which do not apply to the work at hand. The writer must be sure that omission of such parts does not distort the meaning of the original. To indicate to the reader that an element in the original has been left out of the quotation, ellipsis marks consisting of three periods . . . are used. If the portion deleted contains a period, it is added to the ellipsis marks . . . ., making four periods. This is to indicate to the reader that what has been omitted was the final portion of a sentence.

DOCUMENTATION

Documentation means substantiation or evidence, in the form of footnotes. Following a citation in the body of a report, the writer places an arabic number one-half space above the line. The number always appears at the end of the reference, and it refers the reader to the bottom of the typed page on which it appears, where the footnotes are numerically arranged. Footnotes are always single-spaced, but double spaces are used between notes. Each footnote is indented to conform to paragraph indentation on that page. As with the number to which it corresponds in the text, the footnote number is raised one-half space above the line. Footnote numbers in the text and at the foot of the page are not punctuated or placed within parentheses.

While the main use of footnotes is in identifying and acknowledging a source of information used in the text, notes are also used to provide additional material for the reader which the author may not wish to use in the body of the report. Notes may also provide additional, supplementary information which would not properly appear in the body of the report but which is nonetheless helpful.

Sources of information cited may be books, journals or magazine articles, reports, newspapers, government bulletins or other publications, public documents, legal references, unpublished dissertations, unpublished correspondence, or interviews. Each kind of source requires its own form in the footnotes—one form for the first time it is cited, and an abbreviated or shortened form if it is again cited or referred to. One must remember that footnotes are provided for the use of the reader who wants to know the source of the information the writer is providing. In each instance footnote information must be given in a form that makes it easily possible for the reader himself to obtain the original source for corroboration.

The basic information in the first reference to a particular source consists of identification of the author, the full title of the work referred to, the specific article or chapter (if this is applicable), the facts of publication, and specific reference to the page or pages within the source. The final authority for the form or spelling of the author's name is always the title page of his book or the way in which the article is signed. This indicates the proper spelling and whether or not the full name or initials should be used. In a footnote the author's first name or initials precede the surname. Following are examples of various ways in which author-information may be given initially in footnotes:

One author
[1] D. H. Lawrence, *Title* (Facts of Publication), page reference.

Two authors
[2] E. M. Forster and Virginia Woolf, *Title* (Facts of Publication), page reference.

Three authors
[3] John L. Rowe, M. L. Rosenblum, and Mary B. Wheat, *Title* (Facts of Publication), page reference.

More than three authors
[4] William M. Hays *et al.*, *Title* (Facts of Publication), page reference.

or

[4] William M. Hays and others, *Title* (Facts of Publication), page reference.

Author is an association
[5] Society of Technical Writers and Publishers, *Title* (Facts of Publication), page reference.

Author is a committee
[6] Committee on Education, American Chemical Society, *Title* (Facts of Publication), page reference.

Author is a public body
[7] U.S. Congress, House of Representatives, *Title* (Facts of Publication), page reference.

[7] California Department of Taxation, Division, *Title* (Facts of Publication), page reference.

[7] Southampton Township, Division, *Title* (Facts of Publication), page reference.

No author, but an editor
[8] Brian D. Banks (ed.), *Title* (Facts of Publication), page reference.

No author
[9] *Title* (Facts of Publication), page reference.

[9] Anonymous, *Title* (Facts of Publication), page reference.

As with the names of authors, titles of publications are given exactly in the form in which they appear in the source. Articles from periodicals or chapters from books are placed within quotation marks. Titles of books, bulletins, periodicals, newspapers, published reports, lectures and proceedings, encyclopedic works, legislative acts and bills, and unpublished theses and dissertations are underlined (italicized). Following are examples of various ways title-information may be given initially in footnotes:

Subdivisions of publications, such as articles in periodicals
[10]Author, "Careers for the Technician," *American Technical World*, XIV (date), page reference.

[11]"Isaac Newton," *Encyclopaedia Britannica*, VII, edition, page reference.

[12]Author, "How to Develop the Report," *Basic Technical Writing* (Facts of Publication), page reference.

[13]World Health Organization, *Bulletin 86* (Facts of Publication), page reference.

[14]Author, "Servicing the Computer," *Electronics World*, Vol. 37, No. 5 (date), page reference.

Names of frequently used professional journals and periodicals are usually abbreviated in footnotes and bibliographic listings. This is to save time, space, and costs. Abbreviations for the scholarly, professional, and scientific journals have become standard. Each technician will gradually become familiar with these in his own field. Full listing and information regarding abbreviations for journals can be obtained in the library. The form of abbreviation used in the various indexes to periodical literature may be one source of information. But the best form to follow is that used in articles in the technical journals covering the technician's own field.

Newspapers
[15]Author (if any), "Growth of Space Industry Watched by Wall Street," *New York Times* (date), page reference.

[16]*New York Rev. Stat.* (date), c. 25, Paragraphs 276, 277.

Unpublished theses and dissertations
[17]Author, *Time and Motion Study of a Computer Center* (unpublished Ph.D. dissertation, University of Oregon, 1959), page reference.

Separately published reports, lectures, proceedings, etc.
[18]Author, "The Writing of Abstracts," *IRE Transactions on Engineering Writing and Speech* (Facts of Publication), page reference.

Encyclopedia works

[19] Author (if any), "Japanese Industry," *Encyclopedia Americana*, VIII, edition, page reference.

Facts of publication of books are enclosed in parentheses in the footnotes and followed by a comma. Depending on the particular circumstances, these facts may include the total number of volumes, number or name of the edition (if other than the first), name of series of which reference is a part, place of publication, name of publisher, and date of publication. Following are examples of various ways publication information may be given initially in footnotes:

[20] Author, *Title* (12 vols., 3d rev. ed., Columbia Classics; Cambridge, Mass.: The Criterion Press, 1937), volume number, page reference.

[21] Author, *Title* (2d ed.; San Francisco: Globe Publishers, Inc., 1951), page reference.

[22] Author, *Title* (3d rev. ed.; Chicago, 1958), page reference.

[23] Author, *Title* (3 vols.; Northwest Guild Association, 1947), page reference.

[24] Author, *Title* (2d ed., 1907), page reference.

Volume numbers of periodicals are given as either roman numerals or arabic numbers, depending on the practice adopted by the publisher and shown on the journal. Also included in footnotes referring to periodicals are the month and year of publication, or the year of publication alone, followed by a comma. Examples are as follows:

[25] Author, "Title of Article," *Name of Periodical*, XV (May, 1951), page reference.

[26] Author, "Title of Article," *Name of Periodical*, 15 (April, 1969), page reference.

How a specific reference within a particular source—book or periodical—may be cited is shown in the following examples:

[27] Author, "Title of Article," *Name of Journal*, XIII (March, 1967), p. 31.

[28] Author, "Title of Article," *Name of Journal*, 13 (January, 1968), pps. 87–94.
[29] Author, *Title* (The Peerless Press, 1970), II, p. 173.
[30] Author, *Title* (Boston: Chowder and Co., Inc., 1939–53), III, p. 239.
[31] Author, *Title* III (London: Queens Press, 1942), p. 14.
[32] Author, *Title* (Edinburgh, 1880–1887), III, Part II, p. 7.
[33] Author, *Title*, Vol. II, Book III, chap. 19.

Following are examples of some of the different forms in which references to pages may be made in footnotes:

Reference to a single page: p. 5.

Reference continues from a given page to the following page: pp. 7 f.

Reference continues from a given page to pages that follow: pp. 10 ff.

Reference to successive pages: pp. 4–29.

Shortened footnotes for subsequent citation. After a reference has been cited initially in full detail, a much shorter form may be used in referring to it again. In years past, only Latin terms and abbreviations were used in footnoting, but this practice has largely given way to the use of English or Anglicized forms in American scholarly and technical reporting and publishing. Still, since some writers prefer the Latin to the English forms, and because they appear in so many of the older published works, a writer must become familiar with them in the event he encounters them or wishes to use them.

The most common of these is the Latin term *Ibid.*, which should be capitalized, underscored, and followed by a period. It is used when a footnote refers to the same work as that immediately preceding it, without intervening notes. *Ibid.* is an abbreviation for *ibidem*, which means "in the same place" or "from the same work."

[1] John C. Hodges, *Harbrace College Handbook* (5th ed.; New York: Harcourt, Brace & World, Inc., 1962), p. 247.
[2] *Ibid.* (identical with previous reference)

The same situation could be cited in the following manner according to current practices:

[1] John C. Hodges, *Harbrace College Handbook* (5th ed.; New York: Harcourt, Brace & World, Inc., 1962), p. 247.
[2] Hodges, p. 247.

      or

[2] *Harbrace*, p. 247.

Either the short form of the author's name or book title is acceptable.

The same choices are present with the formerly familiar Latin terms *op.cit.*, "in the work cited," and *loc.cit.*, "in the place cited." Formerly, these terms were regularly used in footnotes referring to a previously cited reference (*op.cit.*) where an intervening note to another work had occurred. *Loc.cit.* was used in the same situation, except that both the work and the cited pages were the same:

[1] John C. Hodges, *Harbrace College Handbook* (5th ed.; New York: Harcourt, Brace & World, Inc., 1962), p. 247.
[2] *Ibid.* (same source and page as above)
[3] William E. Buckler and Arnold B. Sklare, *Essentials of Rhetoric* (New York: The Macmillan Company, 1966), pp. 67–70.
[4] Hodges, *op.cit.*, p. 187. (same work as in notes 1 and 2, but different page)
[5] Buckler and Sklare, *loc.cit.* (same work and same pages as in note 3)

If the English rather than the Latin forms were used in preceding notes 2, 4, and 5 they would appear as:

[2] Hodges, p. 247.
[4] Hodges, p. 187.

      or

[4] *Harbrace*, p. 187.
[5] Buckler and Sklare, pp. 67–70.

Unless institutional practice prescribes the specific form of footnoting, the choice is an option of the writer. Whichever form he chooses, however, it is essential for him to be consistent in his practice throughout the report.*

TERMS USED IN FOOTNOTING

Following is a short list of Latin words and expressions—with explanatory comment—that may be helpful to the technical researcher. This list is followed by a second list of abbreviations of mainly English words used in research writing and printing.

Latin Terms in Footnotes

Abbreviation	Word	Definition and commentary
......	ante	before
ca.	circa	about (used with dates, e.g., *ca.* 1906)
et al.	et alibi	and elsewhere
et seq.	et sequens	and the following
ibid.	ibidem	in the same place; from the same work
......	idem	the same; the same as that mentioned above
......	infra	below (should not be substituted for *ibid.* or *op. cit.*)
loc. cit.	loco citato	in the place cited; in the passage last referred to
op. cit.	opere citato	in the work cited
......	passim	everywhere; all through, here and there
......	post	after
......	sic	thus (inserted in brackets within a quotation and after a quoted work or words to indicate that the preceding expression, strange or incorrect as it may be or seem, is exactly quoted)
s.v.	sub verbo	under the word or heading above

*Lest the reader think that we preach what we do not practice, he should also know that owners of copyrighted material frequently specify the credit which must be used if their material is reprinted in a publication offered for sale. Thus several styles of acknowledgement, all different, can be found in this volume. In a technical report consistency can be achieved; in a commercial publication this is not often possible—even though it would be desirable from the reader's standpoint.—Ed.

......	*supra*	above (should not be used in place of *ibid.* or *op. cit.*)	
......	*vide*	see	
viz.	*videlicet*	namely, to wit	

Latin and English Terms in Footnotes

Abbreviation			
Singular	Plural	Word	Definition and commentary
art.	arts.	article(s)	
bk.	bks.	book(s)	If Roman numerals follow this abbreviation, it should be capitalized, e.g., Bk. II; if Arabic numerals follow, the abbreviation is not capitalized, e.g., bk. 2
bull.	......	bulletin	
cop.	......	copyrighted	e.g., cop. 1920
cf.	......	compare	
chap.	chaps.		chapter(s)
col.	cols.	column(s)	
diss.	......	dissertation	
ed.	eds.	edition(s)	e.g., 2d ed.
ed.	eds.	editor(s)	or, edited by
e.g.	......	*exempli gratia*	for example
et al.	......	*et alii*	and others
f.	ff.	following	and following page(s), e.g., pp. 5f. means page 5 and the following page; pp. 5ff. means page 5 and following pages
fig.	figs.	figure(s)	e.g., fig. 2 or Fig. II
i.e.	......	*id est*	that is
illus.	......	illustrated	or, illustration
l.	......	line(s)	
MS	MSS	manuscript(s)	may be written ms. or mss.
n.	......	footnote(s)	e.g., "see n. 5" means to refer to footnote number 5
n.b.	......	*nota bene*	note well; take notice
n.d.	......	no date	
no.	nos.	number(s)	
n.p.	......	no place	used in bibliography
N.S.	......	new series	

p.	pp.	page(s)	
par.	pars.	paragraph(s)	
proc.	......	proceedings	
pt.	pts.	part(s)	e.g., pt. 5 or Pts. V, VI
q.v.	......	*quod vide*	which see; reference is made to it
rev.	......	revise	or, revised, or revision
sec.	secs.	section(s)	
trans.	......	translated	or translation, translator
vol.	vols.	volume(s)	e.g., vol. 4, or Vol. IV, or 6 vols.

BIBLIOGRAPHY

A bibliography is by definition a listing of writings relating to a given subject or author. A technical report which has drawn from diverse printed sources will gather and list its materials in a final bibliography. Technical research and reporting depends for its growth upon accumulated knowledge. Each reporter adds another bit of information to the growing storehouse. One worker adds to and builds upon the information put forward by another. A bibliography of writings relating to the subject of his report can be, for some readers, the most valuable information a technical writer can provide. The reader asks only that the bibliography be as complete as possible, that it be presented in acceptable form, and that the information it contains be correct in all of its detail.

Information in the bibliographic listing for a published source is much the same as that provided in a footnote, but the form is different. Footnotes, of course, assume their order from a relationship with the text of which they are an integral part. Readers consulting footnotes while reading the text are interested in the notes mainly as a source of illumination for what they are reading. By their very nature, items appearing in footnotes have a random order of listing. A bibliography organizes this random listing into a coherent system of greater usefulness to the reader.

Items are listed in a bibliography alphabetically according to authors' surnames. If the author's name is not given, the item is alphabetized according to its title. Following the author's last name, his first name or initials are given. This is followed by the title of the book, the publisher, the place of publication, and the date of publication. Bibliographic entries for journal articles end with the inclusive pages on which the article is found, as in the following examples:

Articles
>Allison, T., "Employee Publications: There's Room for Improvement," *Personnel Journal*, July, 1954, vol. 31, pp. 56–59.
>
>Inglis, John B., "Recent Statements Show New Techniques in Annual Reporting Are Being Widely Used," *Journal of Accounting*, December, 1950, vol. 90, pp. 474–478.
>
>Paterson, Donald G., and James J. Jenkins, "Communication between Management and Employees," *Journal of Applied Psychology*, February, 1948, vol. 32, pp. 71–80.

Bibliographies are often arranged according to subject matter, if the entries run to large numbers. Or the lists may be broken down into categories of publications, such as books, articles, public documents, monographs, etc. In all such cases entries under each category are given alphabetically by the author's last name.

Following is a short, sample bibliography of books on the general subject of technical writing.

>Baker, C., *Technical Publications*, Wiley, New York, 1955.
>
>Baker, J. C. Y., *Guide to Technical Writing*, Pitman, New York, 1961.
>
>Bleckle, Margaret B. & Kenneth W. Houp, *Reports for Science and Industry*, Holt, New York, 1958.
>
>Brown, James, *Casebook for Technical Writers*, Prentice-Hall, Englewood Cliffs, N.J., 1961.
>
>Clarke, Emerson, *How to Prepare Effective Engineering Proposals*, T. W. Publishers, 1962.
>
>Comer, David B. & Ralph R. Spillman, *Modern Technical and Industrial Reports*, Putnam, New York, 1962.
>
>Fountain, A. M. & others, *Manual of Technical Writing*, Scott, Foresman, Chicago, 1957.
>
>Godfrey, James W. & Geoffry Parr, *Technical Writer*, Wiley, New York, 1959.
>
>Hicks, Tyler G., *Successful Technical Writing*, McGraw-Hill, New York, 1959.
>
>_____, *Writing for Engineering and Science*, McGraw-Hill, New York, 1961.

Kelly, R. A., *The Use of English for Technical Students*, Harrap, London, 1962.

Mandel, Siegfried & David L. Caldwell, *Proposal and Inquiry Writing: Analysis, Techniques and Practice*, Macmillan, New York, 1962.

Marder, Daniel, *Craft of Technical Writing*, Macmillan, New York, 1960.

Menzel, Donald H., Howard Mumford Jones & Lyle G. Boyd, *Writing a Technical Paper*, McGraw-Hill, New York, 1961.

Miller, Walter J. & Leo C. A. Saidla, *Engineers as Writers*, Van Nostrand, Princeton, N.J., 1953.

Mills, Gordon H. & John A. Walter, *Technical Writing*, Holt, New York, 1954.

Mitchell, John, *Handbook of Technical Communication*, Prentice-Hall, Englewood Cliffs, N.J., 1962.

Nelson, J. Raleigh, *Writing the Technical Report*, 3rd ed., McGraw-Hill, New York, 1952.

Piper, H. Dan & Frank E. Davie, *Guide to Technical Reports*, Holt, New York, 1958.

Production and Use of Technical Reports, Fry, B. M. & J. J. Kortindick, eds., Catholic, Washington, D.C., 1955.

Racker, Joseph, *Technical Writing Techniques for Engineers*, Prentice-Hall, Englewood Cliffs, N.J., 1960.

Rathbone, Robert R. & James B. Stone, *Writer's Guide for Engineers and Scientists*, Prentice-Hall, Englewood Cliffs, N.J., 1961.

Schultz, Howard & Robert Webster, *Technical Report Writing: Sourcebook and Manual*, Longmans, New York, 1961.

Sherman, T. A., *Modern Technical Writing*, Prentice-Hall, Englewood Cliffs, N.J., 1961.

Souther, J. N., *Technical Report Writing*, Wiley, New York, 1957.

Style Manual for Technical Writers and Editors, S. J. Reisman, ed., Macmillan, New York, 1962.

Technical Editing, Ben H. Weil, ed., Reinhold, New York, 1958.

Ulman, Joseph N., Jr. & Jay R. Gould, *Technical Reporting*, Holt, New York, 1957.

Weisman, H. M., *Basic Technical Writing*, Merrill, Englewood Cliffs, N.J., 1962.

Welloon, G. P. & others, *Technical Writing*, Houghton Mifflin, Boston, 1961.

Winfrey, Robley, *Technician and His Report Preparation*, Iowa State University Press, Ames, Iowa, 1961.

Zall, Paul M., *Elements of Technical Report Writing*, Harper & Row, New York, 1962.

Chapter 10
VISUAL AIDS

Charts, graphs, tables, diagrams, maps, drawings, and photographs help greatly to communicate information in technical writing. Visual presentation of facts is helpful to the reader and saves time and space for the writer. The purpose of graphic illustrations is to clarify and support the text of the writing. This is done by emphasizing certain details and conclusions, making compact comparisons and analyses, indicating ratios and degrees of change in data, and neatly displaying statistical information.

Discussion in depth of graphic methods is outside the scope of a book on basic technical writing. Preparation of certain technical or engineering drawings, for example, requires special training and experience beyond those of the average technician-writer. Industrial plants requiring special visualizations for their technical literature usually employ an artist or draftsman on a full-time basis or purchase these particular services outside the company. Where such services are required, the job of the writer is to prepare data for the graphic artist and to indicate the nature of the communication problem so that the draftsman may best determine what kind of table or pictorial representation is needed.

In a large sense, anything that pictorially represents information could be considered a graphic aid. In this chapter, however, graphs and tables will be considered in a very limited scope. Chiefly, graphs and tables which present numerical information will be discussed. The hope is to do no more than call attention to the subject of graphics in technical writing, suggest the broad usefulness of visualizations, and review briefly some of the ideas relating to the preparation of the most familiar kinds of graphs and tables. Technically-trained students usually approach graphics with greater skills in mathematics and measurement than those not so-trained.

GRAPHS

The bar graph. The bar graph consists of a set of uniformly placed horizontal or vertical bars of equal widths, each starting at the zero point of a scale placed parallel to the bars. Spacing between bars is uniform, unless the spaces represent intervals of time, in which case each space is proportional to the time elapsed. Zero is included on the scale which is used, and the identity of each bar is clear.

Bar graphs permit immediate comparison of quantities and sizes, particularly of variables with unconnected measurements—the length of the horizontal or vertical bars depicting values of the variables. Bar graphs emphasize the difference in amounts rather than the fluctuation of something over a period of time. Conclusions may be drawn by comparing the amounts or percentages represented by the lengths of the bars. (*Figures 1 and 3.*)

The line graph. The line graph is usually a broken or curved line placed with reference to a set of scaled axes. The vertical scale generally starts at zero and increases from bottom to top. The horizontal scale increases from left to right and is used for time or some other variable.

Line graphs are used to show trends or cycles, to plot a series of many successive values, to emphasize movement rather than actual amounts, and to identify relationships between variables. Conclusions may be drawn by comparing the amounts or percentages represented by different points on the line with scale values of these points. (*Figures 2 and 5.*)

The circle graph (pie chart). The circle graph is a circle which represents the whole or 100%, divided by radii into sectors which represent component parts of the whole. It is used to show the relationship of relative sizes of component parts to the whole and to each other. The sectors are usually arranged clockwise in descending order of size, with the largest shown at the top. Amounts represented by the areas of two sectors are compared by the lengths of their arcs or the sizes of their central angles. (*Figure 4.*)

GUIDELINES FOR GRAPHS

The principles of preparation and use of graphics are quickly summarized. The reader, first of all, expects a graphic to be well planned and carefully prepared. A visual should be easy to read and interpret, well spaced, and uncluttered. Details must be accurate—for example, when a graphic is scaled, the scaling must be fully correct and clearly indicated. All drawings are fully labeled. Oftentimes, labels may be placed directly on the drawing, depending on the amount of labeling and the size of the drawing. If this is not possible, parts of the drawing are assigned numbers or letters which refer to a key someplace on the same page. Many times, it is possible to connect the letter or number to the key with a line, in the interest of simplifying the reading process.

The reader also expects to find a consistent system of numbering and titling each visual. Numbering visuals as Figures 1, 2, 3, etc. is the most common practice, and care is exercised so that full correlation exists between the figures and reference to them in the text. Each graphic representation is given a clear title. The viewer must know unmistakably from the title what the data represents. The title is brief only to the extent that brevity is consistent with clarity.

Placement of a visualization is determined by the writer on the basis of greatest usefulness to the reader. Perhaps the ideal is to place each graphic within the body of the text as close as possible to the discussion which it accompanies. Obviously, the mechanics of typing and printing do not frequently allow this, and a compromise is necessary. Some technical writers prefer to collect all visualizations in a separate section following the text or inserted about halfway through it, to facilitate forward and backward reference. Whichever system is used, the need for correlation between references in the text and numbering and titling of drawings becomes even more evident. (*Figures 5, 6, 7, and 8.*)

STATISTICAL TABLES

A table is used to present large amounts of data in a precise and orderly fashion. The most effective table is one which systematically lists relevant data in minimum space.

Characteristically, a table presents information compactly in columns and rows. Brief descriptive headings and subheadings clearly label each column, or easily understood abbreviations may be used. It is permissible to use footnotes to clarify particular items in the table. If special directions are required to interpret the table, these usually appear beneath the presentation. A descriptive title, of course, and a table number always appear above the table. To make use of the table easier for the reader, totals or other subdivisions of the data are usually set off by means of a double-ruled or bold horizontal line. (*Figure 9.*)

COLLECTING STATISTICAL DATA

There are many sources for statistical data, but local, county, state, or the federal government may be the chief suppliers of numerical information required for technical writing. Public and college libraries usually have on hand the following statistical reports published by federal agencies: (all may be obtained from the Superintendent of Documents, U.S. Government Printing Office, Washington, D.C.)

The Decennial Census of the United States
The Monthly Labor Review
Federal Reserve Bulletin
Survey of Current Business
Statistics of Income
Handbook of Labor Statistics
Agricultural Statistics
Statistical Abstract of the United States
Historical Statistics of the United States

Two sources of data for tables not published by the government are *The World Almanac* and *The Economic Almanac*.

CHARTS

Flow chart. A flow chart is used to present vividly the steps in the sequence of a process. The chronologic sequential happenings may be given a sense of movement when seen visually in the process of flow. Such a chart is usually not intended to provide precise details of a process, such as time schedules or job instructions. Flow charts are chiefly valuable for a broad overview of the separate orderly steps which, when integrated, comprise the completed process. Pictorial drawings and cartoons are commonly used in flow charts to heighten interest, stress differentiation of function or responsibility, and to communicate more effectively with the nontechnical, general reader. (*Figures 10 and 13.*)

Organization chart. An organization chart is an effective visual device for showing the arrangement of an enterprise in terms of lines of authority, lines of coordination and liaison, titles of positions, and levels of authority. Lines of authority are usually depicted in heavier ink than the rest of the chart. Broken lines represent lines of consultation, liaison, or coordination. Blocks on the same level usually suggest the same level of authority. In a very large company or organization, several charts are often necessary to show the complexity of the structure. The master chart will often show the top management or corporate structure, while a separate chart will be used for a single division or entity. Sometimes, names of individuals will appear in the blocks, but titles alone is the most common practice. Lines of authority, levels of responsibility, and kinds of working relationships are the essential things in any such chart. (*Figure 11.*)

Statistical map chart. A statistical map chart or cartogram is a most helpful visual means of showing how a variable distributes within or over a geographical area. Map outlines of the state, county, or entire country are available for the grapher from art supply companies; or such outlines can be made from standard road maps, public documents, or atlases. To present the relationship of numbers to geographic locale, coloring, shading, numbers, percentage figures, dots, lines, or pictorial symbols may be shown—whichever is most feasible and likely to be clearest. The purpose of a statistical map chart always remains the same—that is, to show visually the overall proportional numerical relationships in terms of topographic distribution. (*Figure 12.*)

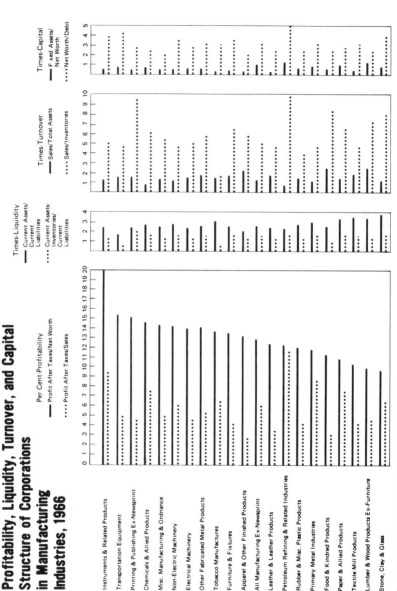

FIGURE 1
Horizontal Bar Graph

*Reproduced by permission of National Industrial Conference Board, New York. *A Chart Guide to Financial Markets*, © NICB, Inc.

VISUAL AIDS

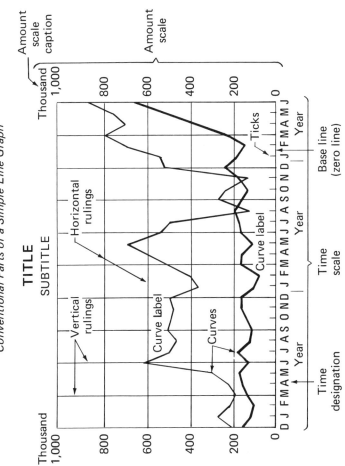

FIGURE 2
Conventional Parts of a Simple Line Graph

FIGURE 3
Vertical Bar and Line Graphs

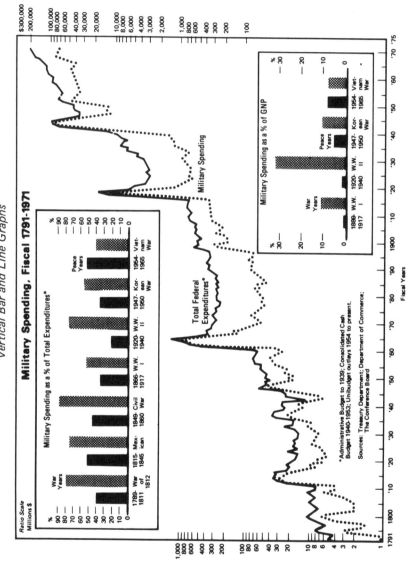

**Reproduced by permission of National Industrial Conference Board, New York. *The Federal Budget, Its Impact on the Economy, Fiscal 1971 Edition.* © NICB, Inc.

VISUAL AIDS 147

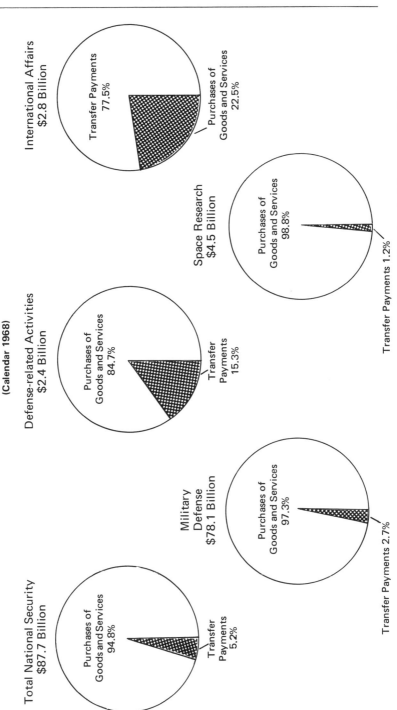

FIGURE 4
Circle Graphs
NATIONAL SECURITY: PURCHASES AND TRANSFER PAYMENTS
(Calendar 1968)

*Reproduced by permission of National Industrial Conference Board, New York. *The Federal Budget, Its Impact on the Economy, Fiscal 1971 Edition.* © NICB, Inc.

Sources: Department of Commerce; The Conference Board

FIGURE 5

Line Graph with Surface Shading and Accompanying Explanatory Text

Persons by Educational Level

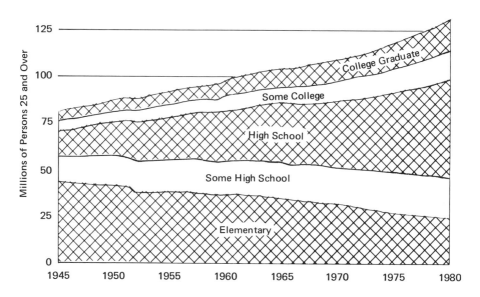

The Degrees of Learning

The rising tide of educational attainment is making for a sizable population of relatively well schooled citizens. At the end of the war, for example, less than 10 million persons had some college training, but by 1960 their number had increased to almost 17 million, and by the end of the decade to almost 20 million. By 1980 close to 33 million Americans will have had some college training, and more than half will have a B.A. degree. Similarly, the number of adults with at least a high school diploma will grow in number from roughly 35 million at the close of the Sixties to more than 50 million by the end of the Seventies.

Thus, in the coming decade there will be a large increase in the relatively well educated segments of the consumer population. Persons with modest schooling, those who have not completed high school, will decline in number from over 50 million at the end of the Sixties to 47 million in 1980.*

*Reproduced by permission of National Industrial Conference Board, New York. *The Consumer of the Seventies.* © NICB, Inc.

FIGURE 6

Modified Bar Graph with Accompanying Explanatory Text

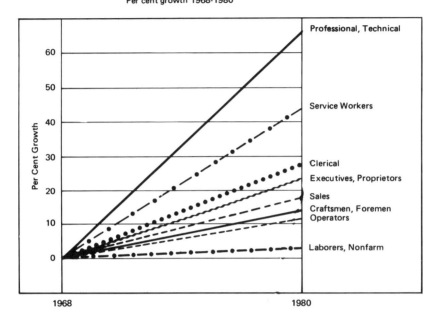

Growth in the Labor Force by Occupation
Per cent growth 1968-1980

Blue Collar, White Collar

The occupational mix of the nation's labor force is changing. The rapid growth of technical and scientific industries, computerization, and in general the growing administrative complexity of the modern business enterprise have intensified the need for white collar employees. At the same time, rising productivity in manufacturing has made possible an imposing increase in physical output with a relatively modest increment in the number of workers required on the assembly line. At the end of the war blue collar workers were some 15% more numerous than white collar workers, but by 1955 they were about equal in number.

In the final years of the Sixties there were some 30% more persons employed in white collar pursuits than in blue collar jobs, and by 1980 the difference will widen to an estimated 60%. In that year half of all employed persons will be in white collar jobs; immediately after the war the ratio was one third.*

*Reproduced by permission of National Industrial Conference Board, New York. *The Consumer of the Seventies.* © NICB, Inc.

FIGURE 7
Vertical Bar and Line Graphs

Capital Structure of Manufacturing Corporations

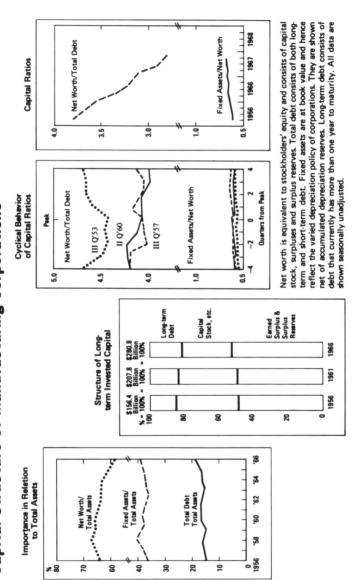

Source: Securities and Exchange Commission; Federal Trade Commission

*Reproduced by permission of National Industrial Conference Board, New York. A Chart Guide to Financial Markets. © NICB, Inc.

VISUAL AIDS 151

FIGURE 8

Statistical Information Adapted to Three Kinds of Graphs

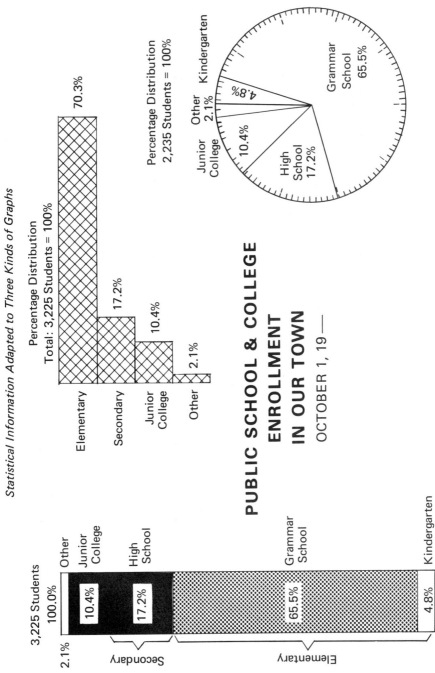

FIGURE 9

Table of Comparative Statistical Data

Total Employment Cost — Wage Employees in the Iron and Steel Industry

Year	Average No. of Wage Employees	Average Hours Worked Per Week Per Employee	Pay for Hours Worked — Straight Time Regular	Pay for Hours Worked — Straight Time Shift Differentials	Pay for Hours Worked — Sunday Premium	Pay for Hours Worked — Overtime Premium	Pay for Hours Worked — Premium for Work on Holidays	Pay for Hours Worked — Sub-Total Pay for Hours Worked	Comparison Average Hourly Earnings in Steel (BLS)	Other Payroll Costs — Holidays Not Worked	Other Payroll Costs — Vacation	Other Payroll Costs — Adjustments	Total Payroll Cost	Employee Benefits Cost Per Hour	Total Employment Cost Per Hour	Aggregate Payroll Wage Employees (Thousands)	Year
	(1)	(2)	(3)	(4)	(5)	(6)	(7)	(8)	(9)	(10)	(11)	(12)	(13)	(14)	(15)	(16)	
1969	415,301	38.7	$3.776	$.050	$.072	$.154	$.068	$4.120	$4.09	$.053	$.370	$.023	$4.566	$.809	$5.375	3,813,449	1969
1968	420,684	37.7	3.553	.049	.067	.126	.065	3.860	3.82	.049	.376	.018	4.303	.729	5.032	3,570,606	1968
1967	424,153	37.0	3.398	.048	.063	.091	.056	3.656	3.62	.056	.349	.008	4.069	.689	4.758	3,330,529	1967
1966	446,712	38.2	3.353	.049	.065	.110	.059	3.636	3.58	.049	.324	.008	4.017	.616	4.633	3,571,600	1966
1965	458,539	37.7	3.261	.049	.064	.106	.056	3.536	3.46	.048	.343	.008	3.935	.540	4.475	3,549,102	1965
1964	434,654	38.3	3.177	.049	.061	.090	.054	3.431	3.41	.048	.310	.007	3.796	.559	4.355	3,300,516	1964
1963	405,536	37.4	3.166	.047	.055	.069	.053	3.390	3.36	.050	.240	.007	3.687	.560	4.247	2,913,214	1963
1962	402,562	36.6	3.141	.046	.052	.052	.041	3.332	3.29	.058	.224	.008	3.622	.533	4.155	2,783,784	1962
1961	405,924	36.6	3.054	.045	.050	.048	.044	3.241	3.20	.055	.200	.005	3.501	.488	3.989	2,713,735	1961
1960	449,888	35.7	2.916	.046	.050	.045	.037	3.094	3.08	.054	.195	.006	3.349	.471	3.820	2,813,992	1960
1959	399,738	36.9	2.896	.048	.060	.082	.058	3.144	3.10	.047	.220	.006	3.417	.381	3.798	2,628,339	1959
1958	411,565	35.2	2.787	.038	.040	.031	.035	2.931	2.91	.056	.190	.004	3.181	.332	3.513	2,405,995	1958
1957	508,434	37.2	2.582	.035	.029	.047	.036	2.729	2.73	.038	.147	.003	2.917	.299	3.216	2,877,819	1957
1956	509,231	38.6	2.407	.036	.008	.066	.025	2.542	2.57	.028	.126	.004	2.700	.254	2.954	2,772,911	1956
1955	519,145	39.2	2.246	.036	—	.065	.029	2.376	2.41	.023	.105	.005	2.509	.213	2.722	2,665,411	1955
1950	503,309	39.0	1.603	.023	—	.055	—	1.681	1.691	—	.064	.001	1.746	.162	1.908	1,785,910	1950
Increase 1969 vs 1950	−88,008	—	$2,173	$.027	$.072	$.099	$.068	$2,439	$2.399	$.053	$.306	$.022	$2,820	$.647	$3,467	2,027,539	Increase 1969 vs 1950
% Increase 1969 vs 1950	−17.5%	—	135.6%	117.4%	—	180.0%	—	145.1%	141.9%	—	478.1%	—	161.5%	399.4%	181.7%	113.5%	% Increase 1969 vs 1950

NOTES FOR TABLE ABOVE by column numbers

(1) Average of the monthly number of employees receiving pay after adjustment for turnover.

(3) Includes cost of living adjustment (effective January 1957 through August 1965 and which was incorporated into the base rates effective September 1, 1965) and incentives.

(5) Sunday premium pay was initiated September 1, 1956.

(7) and (10) Prior to 1953, cost of pay for holidays and premium for holidays worked were included in overtime premium. Pay for holidays not worked was instituted in 1952.

(9) BLS average hourly earnings for 1965 do not include the one-time payment made by most companies in late 1965 resulting from the Extension Agreement covering the period May 1 through August 31, 1965, which provided for an employment cost accrual of 11.5¢ per hour worked.

(11) Includes regular vacation and company liability for vacation time-off and/or pay in lieu of SVP. Appropriate adjustments have been made beginning with the year 1963 in columns 11, 13, 14 and 16.

(12) Adjustments include retroactive payments and adjustments for prior periods.

(13) and (16) Total payroll of wage employees in column 16 is the aggregate annual amount of total payroll cost shown in column 13, and does not include employee benefits cost shown in column 14.

(14) Pension, Insurance, F.I.C.A. taxes, unemployment compensation taxes, and state disability taxes estimated for years prior to 1950. Supplemental unemployment benefits and savings and vacation plan benefits were initiated August 3, 1956 and July 1, 1962, respectively. (Only that portion of Savings and Vacation plan benefits not included in payroll is reported in column 14.)

WAGE RATE INCREASES AND MINIMUM WAGE RATES IN THE IRON AND STEEL INDUSTRY 1965 – 1969**

Effective Date	Wage Rate Increases Per Hour			Minimum Std. Hourly Wage Rate (Job Class 1 and 2)
	Job Class 1 and 2	Increment/ Job Class	Avg. All Job Classes	
August 3, 1956	$.075	$.003	$.095	$1.820
January 1, 1957	.030	—	.030	1.850
July 1, 1957	.110	.002	.122	1.960
January 1, 1958	.050	—	.050	2.010
July 1, 1958	.110	.002	.122	2.120
January 1, 1959	.010	—	.010	2.130
December 1, 1960	.070	.002	.088	2.200
October 1, 1961	.085	.001	.092	2.285
September 1, 1965	.100	.003	.121	2.385
August 1, 1967	.060	.005	.076	2.445
August 1, 1968	.200	.005	.236	2.645
August 1, 1969	.120	.003	.142	2.765

**Note As of August 1, 1969, the standard hourly wage rate varies from a minimum (Job Class 1 and 2) of $2.765 to a maximum (Job Class 33) of $5.338 with an increment of 8.3¢ per hour between job classes.

*Reproduced by permission of American Iron and Steel Institute, Washington, D.C.

VISUAL AIDS 153

FIGURE 10
An Industrial Flow Chart

FLOW CHART OF STEELMAKING

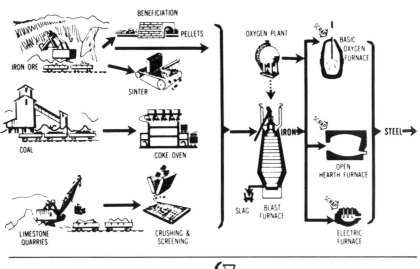

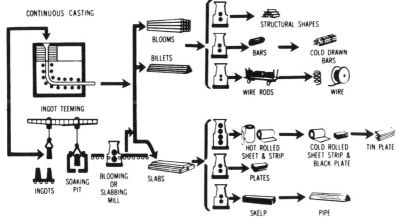

*Reproduced with permission of American Iron and Steel Institute, Washington, D.C.

FIGURE 11
Organizational Chart

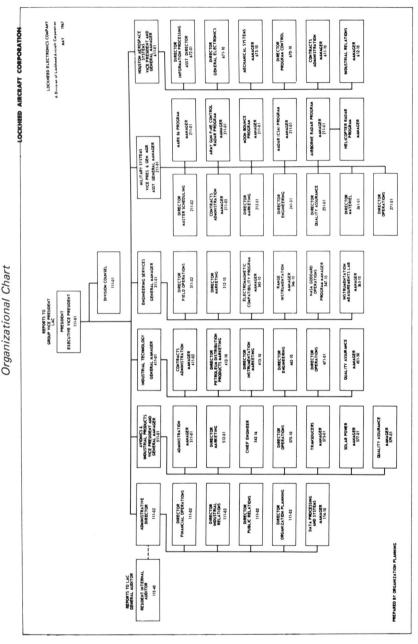

*Reproduced by permission of National Industrial Conference Board, New York. *Corporate Organization Structures, Studies in Personnel Policy No. 210.* © NICB, Inc.

VISUAL AIDS

FIGURE 12
Statistical Map Chart

Net Federal Budget Impact By States

(Federal Payments Per Dollar of Revenue, Fiscal 1965-1967)

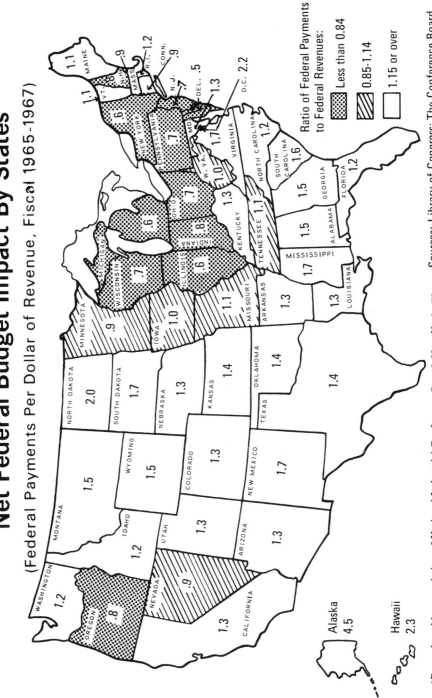

*Reproduced by permission of National Industrial Conference Board, New York. *The Federal Budget, Its Impact on the Economy, Fiscal 1971 Edition.* © NICB, Inc.

Sources: Library of Congress; The Conference Board

FIGURE 13
Visual Explanation of Technical Process

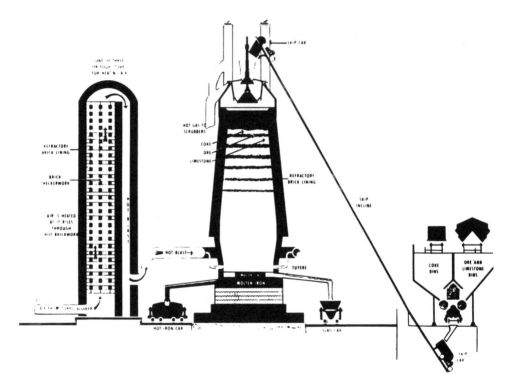

CHARGING THE FURNACE — Iron ore, coke and limestone are kept in storage bins (at right below) and are fed into the furnace. These raw materials drop from the bins into a scale car and are carried to a skip car, which travels up the incline to dump its load in the hopper. The materials melt in the intense heat, settling and decreasing in volume. The skip cars dump in fresh loads at a steady rate, to keep the furnace full.

HEATING THE FURNACE—Air enters bottom of stove, at left. It rises through heated bricks arranged in an open, checkerboard pattern. The air picks up heat from the bricks, reaching 2000 degrees F or more. The hot blast enters the furnace through pipes called tuyeres.

TAPPING THE FURNACE—Periodically the slag which forms on top of the molten iron is drawn off into a slag car. The iron flows into a hot metal car, also called a bottle car. Furnace runs continuously until repairs are needed or the demand for its output falls off.

*Reproduced by permission of American Iron and Steel Institute, Washington, D.C.

Part II
THE TECHNICIAN WRITES LETTERS, MEMOS, ARTICLES, ABSTRACTS

Chapter 11
THE BUSINESS LETTER

While there are many modes of communication in the contemporary world of technology, science, industry, business, and government, the business letter remains the most common and familiar form. Technical personnel may discover that letter writing is one of their chief activities.

A business letter deals with a single specific item of mutual concern to sender and receiver. As a representative of a particular company, however, the writer will always wish to help build good will for his agency through the communication. If a letter creates a poor impression, it is a poor impression of both the writer and the organization he represents.

Since the letter continues to be a main staple in the activities of modern business and industry, many companies have developed their own manuals and guidebooks for letter-writing. There will be variations and modifications in the form of the letter according to the particular needs of the specific company or agency. Yet the broad purposes which apply generally to all business letters are recognized to be the same. Each letter communicates a message dealing with a specific item of business while at the same time providing a permanent record of the exchange or transaction. Each letter serves essentially as a bridge of understanding between the writer and his organization, on one side, and the receiver, on the other.

PARTS OF THE BUSINESS LETTER

Traditionally the business letter has consisted of 6 parts: *the heading, the inside address, the salutation, the body, the complimentary close,* and *the signature.*

The heading The heading indicates who the letter is from—what company—and when it was written. The *who* is of course supplied by the company's letterhead, which gives not only the full corporate title of the organization but its complete address, and, more often than not, the phone number and/or cable address, where appropriate. Most headings not only indicate clearly the *who* and *where* of the writer—they show *what* the company does. A brief descriptive phrase often appears someplace on the letterhead to describe the nature of the business the company engages in, whether manufacturing ball bearings or providing a computer analysis service.

The date is typed in below the company name and address in the position determined by company practice. It is usually either centered or toward the right margin, several spaces below the engraved or printed letterheading.

Clearly, since the company designs and provides the stationery, the heading offers no difficulty to the technician.

The inside address. While the heading tells the who and where of the sender, the inside address indicates the name and address of the person and firm of the recipient. This is noted just as it will appear on the envelope, including the full title of the receiver as well as the postal zone number. Basically, the practice of including the inside address in this form is a matter of routine office procedure for convenience in addressing envelopes and filing carbon copies. The inside address is placed at the left margin, at least two spaces below the date.

The salutation. At the left margin, two spaces below the address, is the salutation, the words with which the writer greets the person to whom he is writing. The standard greeting in the male singular is *Dear Sir*; the plural is *Gentlemen*. In the feminine, *Dear Madam* and *Ladies* are used in the singular and plural. However, a most pronounced trend in recent decades has been to greet the recipient personally by his own name and title: *Dear Dean Smith, Dear President Jones, Dear Mr. Johnson, Dear Mrs. Schultz*. This practice is recommended in the interest of making letters less formal and more personal.

In general, the greeting must agree with the inside address in number and gender. If the inside address indicates *Mr. Jack Jones, Packaging Manager*, the greeting may not be to *Gentlemen*. It must be to *Dear Sir* or to *Dear Mr. Jones*. The greeting is always punctuated with a colon–nothing more.

The body. The body of the business letter begins two spaces below the greeting. The lines of type within a paragraph are single spaced, but double spacing is used between paragraphs. More will be said about the body of the letter in the discussion which follows on the style and mechanics of letter writing.

The complimentary close. At least two spaces after the last line of the body of the letter, usually beginning at the center of the page or some spaces to the right of it, the *complimentary close* or sign-off appears. Like the salutation, the complimentary close is standard. The acceptable forms most widely used are: *Very truly yours, Yours very truly, Yours truly,* and *Truly yours; Sincerely yours, Yours sincerely, Yours very sincerely,* and *Sincerely; Cordially yours, Yours cordially,* and *Cordially. Respectfully, Respectfully yours,* and *Yours respectfully* are used where warranted.

The signature. The signature of a standard business letter appears directly below the complimentary close and consists of several parts, depending on the practice of the individual company. First may appear the company name, all in upper case letters. Immediately beneath this may come the signed or written signature of the writer. This is followed by the typed name of the writer and his official title. The element most frequently omitted is the first, the name of the company. When all the elements are included, however, the entire signature would appear as:

FORWARD PRECISION COMPANY

R. Clive Brown
R. Clive Brown,
Technical Analyst

or
>   DYNAMIC CHEMICAL CORPORATION
>   *James Blataat*
>   James Blataat,
>   Product Information Technologist

Beneath the signature, at the left margin, for purposes of information and record-keeping, appear the initials of the writer and the initials of the typist. The forms may vary: *TCB/mb; tcb:mb;* or others. If the letter contains enclosures to the recipient, the word *Enclosure* appears beneath the identifying initials.

ATTENTION LINE AND SUBJECT LINE

For a variety of reasons, a standard business letter may be addressed to a company generally but directed to the attention of a specific person within that company. The writer knows, from prior experience, that a particular person at the receiving end is familiar with the business at hand or holds responsibility for a certain function. In such cases the attention line or device is used between the inside address and the salutation. It will appear as:

> Veracity Data Programs,
> 849 Third Ave.,
> New York, New York 10036
>
> Attention of Miss Ruth Brown, Executive Assistant
>
> Gentlemen:
>   or
> Dunhill Industries, Inc.
> 3717 Lawrence Ave.,
> Chicago, Illinois 60613
>
> Attention: Mr. T. C. Croup, Shipping Dept.
>
> Gentlemen:

The attention line may appear either at the left margin, as shown, or it may be centered. In either case, since it may properly be considered an integral part of the inside address, the attention line should also be noted on the envelope of the letter to expedite handling at the receiving end.

The subject line in a business letter is a useful device which helps reduce the need for providing a good deal of information in the opening paragraph of the letter. The reader of the letter is quickly able to determine the nature of the business at hand.

The subject line usually appears in the middle of the page on line with the salutation, but it may also appear at the right margin, on line with the last line of the inside address. Practices vary slightly.

Nirvana Plastics Company,
30 Overhill Circle,
Pittsburgh, Pa. 12345   Subject: Your shipment of Jan. 17, 19——.

Gentlemen:

or

Valhalla Steel Products, Inc.
147 Market St.,
San Francisco, California 12345

Gentlemen: Subject: Your invoice of March 22, 19——.

The most common term used to indicate the subject of a letter is *RE*, followed by a colon. The word comes from the Latin *res* meaning *thing*. In business correspondence the abbreviation of *res*, that is, *Re*, has come by practice to mean *in reference to the following thing or subject*. Some users have come to employ *re* rather than *Re*. In either case the colon is required. The choice in positioning the word is the same as that for the subject line above.

PATTERNS OF INDENTATION AND PUNCTUATION

Though modifications and variations are common, there are generally two different forms of indentation in current business letters. These are called the *block form* and the *idented form*. The patterns are shown in the following designs.

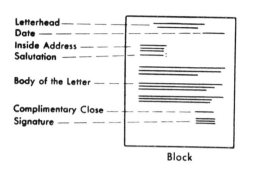
Block

Indented Form

Observation of some of the business letters currently received will indicate that many companies today have adopted features of both the block and indented forms. From the writer's point of view, consistency is necessary in whatever pattern of indentation he or his company uses.

In terms of the overall pattern of the letter, consistency is necessary not only in the form of indentation but in the style of punctuation adopted for the date, inside address, complimentary close, and signature. Writing practice refers to *open* and *closed* puncutation. Closed punctuation calls for a comma at the end of each line and a period at the end of the last line. This would apply to the inside address and the complimentary close, since a colon is always used with the salutation. Open punctuation is clearly the most popular practice today, however, since it requires no punctuation of the lines of the principal parts of the standard letter and therefore saves both time and money. Closed punctuation would appear as:

                                                    February 9, 19—

Mr. James Kindred, President
Utopia Glass Manufacturing Company,
7913 Glistening Pond Road,
Waterloo, Kansas 12345.

Dear Mr. Kindred:

                                                  Yours very truly,

Open punctuation would appear as:

Blasted Dynamite Corporation
1213 Inferno Road
Satania, South Dakota 12345

Gentlemen:

         Sincerely yours
         BOTTOM CHEMICAL COMPANY

         Peter C. Gray, Technical Writer

GUIDELINES FOR THE BUSINESS LETTER

Technical and business English. The old style of business English–"Yours of the 9th inst. at hand and contents duly noted," or "We are in receipt of yours written under the date of February 9th"–has long been abandoned. Standardized conventional phrases and cliches in business correspondence are no longer acceptable. Business English today strives to be natural, simple, direct, and sincere. The writer uses a style which is warm and informal but at the same time courteous and tactful.

 Technical terminology is both appropriate and acceptable in a business letter, provided the receiver of the letter is likely to understand. If any technical terms are used which may prove unfamiliar to the reader, brief definitions are in order. Technicians are frequently obliged, however, to correspond with nontechnical people who may be totally unfamiliar with technical matters. In such instances technical terms would be discourteous and unproductive.

Organization and thoroughness. Each business letter is designed to deal with a specific item of business, a particular subject. It is costly to writer and reader if all aspects of the subject at hand are not covered in the initial correspondence. Each time that additional letters are required with supplementary information, time and money are wasted. The writer is therefore obliged to plan his letter thoroughly beforehand so that all pertinent details will be provided for the correspondent. Even if a plan for a letter consists of no more than a simple list of items to be covered, it is certain to be useful in organizing and covering all aspects of the subject.

The business letter which is well organized is likely to be thorough and explicit. Details relating to specifications and particulars will readily find their place in the organizational pattern. Exact descriptions, methods, and arrangements will not be left hanging for future exchanges. The beginning writer of business letters is advised to remember that a casual attitude that leaves explicitness and thoroughness to chance is likely to produce unsatisfactory results. Organization is the best protection against possible failure.

Diction and tone. Direct, simple, and clear wording is the best. The writer should try to use short, familiar everyday words rather than long words or technical language. While it is neither possible nor desirable to avoid all long or technical language, the writer interested in maximum communication will keep difficult diction to a minimum.

While a business letter is written by one person to another, the writer should not forget that he represents the business that employs him. The best letter conveys a feeling that the writer has a genuine wish to be helpful. The tone is tactful, courteous, and natural. A letter must not antagonize or irritate the reader through a lapse in taste, temper, or tone on the part of the writer.

Examples: Business Letters (Block Form, Open Punctuation)

X-Y-Z COMPANY, INC.
Fielding, Lakeland

October 8, 19__

Professor W. B. White
University of Aquarius
Institute of Astro Sciences
17 Moon Drive
Aquarius, Galaxy 5000

Dear Professor White:

It was a great pleasure to receive the preprint of your paper, "On the Structure of Water and the Possible Existence of Long-Range Order." I had the good fortune to hear your presentation at the Annual Research Conference, although I did not have a chance to speak with you there.

We are tremendously interested in the subject of water in the phenomenon of food preservation, and separately have great interest in the entire subject of slush making. In the latter area, we own or lease more than 1 million acres of land and operate several large mills which produce both reacted and unreacted slush in very large quantitites.

For reasons of industrial secrecy I am not free to discuss our precise interests without a Secrecy Agreement. However, it might be desirable to pursue the possibility of your working with us as a Consultant in either or both of the above areas. In similar cases, we have invited professors to visit us for a day and to present a lecture to our interested staff members, followed by a period of consultation. In such cases, we are pleased to reimburse our guest for his consulting and travel expenses.

It is my personal belief that the general area of the structure of water and slush additives is one of the most interesting in which our company is engaged. Since we have a Technical Center with several hundred research and engineering personnel, this is of great significance.

I hope that we can develop a productive association with you.

Cordially yours,
XYZ COMPANY, INC.

P. M. Landers
Director, Research

PML:fw

X-Y-Z COMPANY, INC.
Fielding, Lakeland

October 14, 19__

Mr. R. F. Jones, Regional Manager
Thomson C. Ranchmont, Inc.
Sunny Court
Blue Brook, Greenfield 6000

Dear Mr. Jones:

Confirming our telephone conversation of October 8, we may be interested in discussing with the appropriate people at T. C. Ranchmont new ideas they may feel are applicable to the container and packaging industry (covering metal, paper and plastics) which could be submitted to us on a non-confidential basis.

We are interested in locating pertinent areas where T.C. Ranchmont may have invested a substantial number of man-years in developing experience in fields which are non-competitive to packaging but which may be related technically to our work.

As a means of narrowing down the infinite variety this may suggest, we have found recent interests in all of the following areas:

1. <u>TNT</u> - the application of high intensity forces to the various operations involved in packaging.

2. <u>Cave arts</u>.

3. <u>Hammers</u> - including means of improving or of nondestructive evaluation of boxes and jars.

4. <u>Computer applications</u>, especially in machine design, industrial information retrieval or the provision of analytical back-up for laboratory work.

5. <u>Steam processing technology</u>, including fabrication, lamination, dehydration methods, use of wet materials, moisturizing of steam, water/air composites, etc.

6. <u>Levers</u> and their applications in inspection or fabrication in packaging structures.

7. <u>Low cost materials</u>, including weeds, slush, etc.

8. <u>Tinted art</u> - color measurement and control.

9. <u>Staples</u> or other joining techniques for thin gauge materials.

10. <u>Paint technology</u> - application of color to thin gauge materials.

11. <u>Money disposal</u>, especially in regard to packaging.

12. <u>Systems research</u> as it might be applied to the synthesis of complete forming systems.

Should you be able to discover any significant matches between this list and the areas of entrepreneurship on your own staff, we would be most pleased to work out a schedule for a meeting in Blue Brook, which you and I tentatively scheduled for November 12.

         Cordially yours,
         XYZ COMPANY, INC.

         P.M. Landers
         Director, Research

PML:fw
cc: R.B. Wheel
 J.G. Smith
 N.S. Boat
 J.E. Book
 E.H. Mill
 D. J. Green

Example: Business Letter (Block Form, Open Punctuation)

      X-Y-Z COMPANY, INC.
       Fielding, Lakeland
            December 5, 19__

Dr. H. B. Wing, President
Guppy University Foundation
P.O. Box X
Lake Zero, Wonderland 7000

Dear Dr. Wing:

As the result of a letter written by you earlier this year to our Corporate Headquarters in Yellow Field, we have conducted an investigation of the desirability of supporting individual fellowships in Physics at Guppy University. At our request, Dr. Neon was kind enough to send us a detailed report describing the science program there.

After careful review of this program, Dr. J.G. Smith, Director of our Science Research Department, has concluded that there are other educational institutions whose programs can be more closely identified with the work we are conducting here. Our company maintains an extensive Technical Center at this address and has been conducting research and development work in a variety of fields for many years.

Accordingly, we must decline to participate in your program at this time. We appreciate the opportunity to review the University's work.

                                    Cordially yours,
                                    XYZ COMPANY, INC.

                                    P.M. Landers
                                    Director, Research

PML:fw
cc: L.M. Neon - Guppy University
    R.B. Wheel
    J.G. Smith

THE TECHNICAL LETTER

The general requirements for the technical letter are the same as those for the business letter in regard to the component parts of heading, inside address, salutation, body, complimentary close, and signature. It might even be most logical to think of the technical letter as a business letter dealing chiefly with technical matters. Conventionally, business letters are thought of as inquiry and reply letters, order and acknowledgement letters, claim and adjustment letters, credit and collection letters, or sales and good will letters. In any of the foregoing kinds of letters technical information may become part of the body of the letter, although the chief intention of the letter may not be to provide technical explanation. The characteristic of the technical letter which distinguishes it from other kinds is its purpose. That purpose is to offer specific, detailed technical data on a particular subject to a selected correspondent.

Because of the distinct purpose of the technical letter and its usually greater length and complexity, it frequently is obliged to take on some of the attributes of a brief report as regards its organization and format. While the traditional trappings of a business letter are there in terms of salutation and complimentary close, the body of the letter may consist of sections and subsections with headings such as those found in a report. There are no hard and fast rules for the format of a technical letter. The usual practice is to follow the procedure for organizing a report by providing an introduction; a materials, methods, and results section; a discussion section; and a conclusion section. But the mode of organization is subject to many variations and is freely modified by individual writers to suit the needs of the situation at hand. What is clearly evident is that no writer may successfully undertake to write a technical letter unless he possesses prior experience and knowledge of both business letter and technical report writing.

Example: Long Technical Business Letter
(Block Form, Open Punctuation)*

 GENERAL FOODS CORPORATION / *250 North Street, White Plains, N.Y. 10602*

March 22, 19__

Hearing Clerk
Department of Health, Education and Welfare
Room 5440
330 Independence Avenue S.W.
Washington, D.C. 20201

Re: 121 CFR Part 128
    Human Foods
    Current Good Manufacturing Practice (Sanitation)
    Manufacture, Processing, Packing or Holding

Dear Sir:

Reference is made to the above proposed regulations which were published in the Federal Register, Vol. 32, No. 242, December 15, 19__ and for which an extension of time for filing comments was granted in the Federal Register, Vol. 33, No. 31, February 14, 19__.

General Foods Corporation is a Delaware corporation with headquarters at 250 North Street, White Plains, New York. It is a manufacturer of a wide variety of food products and has a large number of plants and facilities throughout the country which would be affected by these proposed regulations.

GENERAL CONSIDERATIONS

1. General Foods is in accord with the basic intent of the proposed GMP. However, in order for regulations to be effective, they should be reasonable and realistic so that a manufacturer governed by these regulations can, with reasonable effort, adhere to the regulations. Some aspects of the proposed GMP appear not to follow these criteria. Frequently, the regulations talk in terms of absolutes, which is a contradiction in terms of a rule of reasonableness. In some places, they impose conditions over situations not within the manufacturers' control. Of

---

*Reproduced with permission.

most concern is the fact that they frequently set up requirements which, if literally interpreted, could transform even the best of plants, presently operating under generally recognized good sanitary practices, into plants manufacturing food products in violation of the Federal Food, Drug, and Cosmetic Act. These points will be discussed in more detail in the comments that follow.

2. In the proposed GMP, the regulations are referred to as "Current Good Manufacturing Practice." General Foods strongly recommends that the name of these regulations be changed to "Guidelines for Good Manufacturing Practice." There are several good reasons for this recommendation.

(a) The Word "current" is defined in the dictionary as "in general use or knowledge; prevalent; also, generally accepted; in vogue." If the word "current" as applied to these regulations is to be interpreted in accordance with this definition, it means, in effect, that the proposed regulations are in general use or prevalent in the food industry. While some of the proposed GMP may well be in current use, other aspects of the regulation appear to be desirable goals for the future rather than current practice. If FDA wishes to issue "current" GMP, it would seem necessary for FDA to hold hearings in order to determine what is actually current in the industry and then to issue regulations in accordance with those findings.

(b) These regulations are being promulgated under Section 701 (a) of the Act. There is a serious question of law whether FDA can promulgate substantive regulations under this section, as opposed to interpretive regulations. If FDA proposes to make these regulations substantive rules of law, the breach of which can result in serious penalties, it would again seem incumbent on FDA to hold hearings to establish the factual basis and need for such regulations.

(c) As indicated above, the proposed GMP do not, in many cases, reflect existing conditions or existing practices. If adopted as substantive rules of law, many plants could literally comply only by making major structural changes or by abandoning present facilities and building new ones. General Foods does not believe this is the intent of such regulations. Nevertheless, there is no time element taken into consideration in these proposed GMP. If the proposed GMP, particularly those sections relating to plant construction and equipment, were to apply to future plants and equipment, except where existing facilities represented a health hazard, this concern over the time element could be largely minimized.

For the reasons expressed above, General Foods recommends that the proposed regulations be adopted as interpretive regulations under the heading, "Guidelines for Good Manufacturing Practice." Such action would help resolve the dubious legal status of such regulations, alleviate much of the concern as to the effect of these regulations on existing operations, avoid the likelihood of protracted legal actions challenging the validity of the regulations, and yet offer useful guidelines for orderly adoption of improved manufacturing practices.

3. The proposed GMP apply not only to plants engaged in manufacturing, processing, packaging and labeling, but also storage of finished packaged foods. Many of the provisions in the GMP would not be necessary for warehousing of finished goods. Yet, in effect, warehouses are held under the proposed GMP, to the same standards as a plant which manufactures food. It would be appropriate to exclude warehouses of finished product from the proposed GMP and establish a separate set of guidelines applicable to warehousing situations. In November 1964, FDA set forth the conditions for sanitary protection of stored food products in FDA Publication No. 25. This was republished in part in FDA Papers, Vol. 1 No. 3. It is recommended that FDA set up separate guidelines for warehousing as set forth in its prior publications.

COMMENTS WITH RESPECT TO SPECIFIC PROVISIONS

Assuming that the proposed GMP are adopted as interpretive "Guidelines," General Foods believes that a number of changes should be made in specific sections to make them reasonable and to reflect the realities of some operational problems. Those recommendations are as follows.

4. Section 128.1 (a) defines "adequate" as being "in conformance with local, state, and public health service requirements or recommendations." The word "recommendations" in subsection (a) should be deleted. Its meaning is so indefinite that it offers no real guidance to a manufacturer and makes him potentially subject to arbitrary and capricious interpretations by regulatory officials.

5. Section 128.1 (b) defines "corrosion-resistant material" as one which will maintain its "original surface characteristic" under prolonged use. If interpreted strictly, this definition could rule out certain types of equipment that are quite acceptable and desirable for food use. For example, electro-plated units of equipment are subject to normal wear and may, over a period of prolonged use, lose their original surface

characteristics. However, such equipment can be properly reconditioned to restore the desired surface characteristics.

6. Section 128.1 (c) defines "<u>readily</u> cleanable" as "<u>easily accessible</u> and of such design, material, and finish that residues from or caused by processing operations may be <u>completely</u> removed by normal safe, cleaning methods."

There are several objections to this definition. First, the word "readily" should be deleted from the defined term. There is much equipment in plants which is "cleanable," but not "readily cleanable." The test for GMPs should not be whether equipment is readily cleanable but whether it is cleanable and is, in fact, kept clean.

The same logic applies to the words "easily accessible." Some parts of plant equipment are not easily accessible, but they are accessible for cleaning purposes. The word "easily" should be deleted.

This definition also requires that residues from processing operations be "completely removed." There are some types of equipment in which it is not possible to completely remove all residues, but which are nevertheless safe in that they do not introduce bacterial contamination into the product. The word "completely" should be deleted. If desired, a word such as "satisfactorily" could be substituted in its place.

7. <u>Section 128.1 (d)</u>. The definition of "plant" raises the question of whether office areas in a plant, not related to the actual manufacturing operation, would be included in the definition. Such office areas should not be included in the definition of a "plant."

8. Section 128.3 (a) applies to "the location and grounds surrounding a food plant." The word "location" should be deleted since the location of existing plants with respect to adjoining properties is no longer a matter under the manufacturers control.

9. <u>Section 128.3 (b)</u>. Reference to "location" should be deleted from this paragraph for the reason expressed in comment 8.

10. <u>Section 128.3 (b) (1)</u>. This paragraph provides that "fixtures, ducts, and pipes shall not be suspended over working areas in such a manner that drip or condensate <u>may</u> contaminate foods, raw materials, or equipment." The word "may" as used in this quoted section can be interpreted to mean any possibility of contamination, no matter how remote. Reasonable and workable guidelines should not be based upon a mere remote possibility. They should be aimed at situations where

one could reasonably conclude that there is some likelihood of contamination. Therefore, it is recommended that the word "may" be deleted and the words "is likely to" be substituted in its place.

11. <u>Section 128.3 (b) (2)</u>. This paragraph also contains the word "may." For the reasons expressed in comment 10, the word "may" should be deleted and the words "is likely to" be substituted in its place.

12. <u>Section 128.3 (b) (4)</u>. This paragraph states: "provide adequate ventilation to eliminate objectionable odors and noxious fumes or vapors ****." The word "eliminate" used in this provision is objectionable because it is, again, an absolute. It would be more reasonable to substitute the word "control" in place of the word eliminate. Also, one of the items to be controlled is "objectionable odors." The question raised by this language is "objectionable to whom." It is often impossible to eliminate all odors in processing, some of which may be objectionable to some persons but not to others. A reasonable guideline should not depend upon variable aesthetic considerations. If the guidelines propose to include "odors" as one of the criteria for GMP, it should be "odors which are likely to contribute to contamination."

13. <u>Section 128.4</u>. This section provides that "all plant equipment and utensils should be ***":

(a) "readily cleanable." The word "readily" should be deleted from this provision for the reasons expressed in comment 6.

(b) "smooth." This requirement as applied to all food processing equipment is inappropriate. Some equipment, such as grinding equipment, is designed to have rough surfaces in order to perform its intended function.

(c) "corrosion-resistant." This requirement could be objectionable if interpreted to exclude certain types of equipment as mentioned in comment 5.

(d) "non-absorbent." The requirement that all equipment and utensils be non-absorbent is inappropriate. In some cases, wood is used without any potential for contamination. In other situations, certain types of equipment are specifically designed to be absorbent in order to draw off excess moisture. This requirement should be deleted or clarified.

14. <u>Section 128.5 (a)</u> requires that "hot and cold running water" shall be provided in certain areas. In some processing areas controlled

temperature water is provided rather than separate hot and cold water. This requirement should be amended to permit the use of "hot and cold or controlled temperature" water.

15. Section 128.5 (d) requires that "doors to toilet rooms shall be self-closing and shall not open directly into areas where food is exposed to airborne contamination." Some plants presently do have toilet rooms opening into food processing areas and it would be extremely difficult, in many cases, to change this situation. However, reasonable alternatives are available to eliminate possible airborne contamination. In some cases, double dooring might be a possible alternative. In other cases, it would be possible to set up a negative air pressure in toilet facilities so that all air flow would move from the processing area to the toilet room, thereby eliminating the likelihood of airborne contamination from such rooms. These alternatives should be added to this section.

16. Section 128.5 (e) relating to adequate hand washing facilities, requires the presence of hot and cold running water. Again, the alternative of "controlled temperature water" should be added.

17. Section 128.6 (a). This subsection, while related to repair and maintenance, contains the following sentence: "only such poisonous and toxic materials as are required to maintain sanitary conditions and for sanitization purposes shall be used or stored in the plant." This language, if applied to all materials in the plant, rather than merely cleaning and sanitizing materials, would prevent the storage of materials essential to maintenance and processing, such as certain food additives, lubricants, etc. This language should be clarified so as not to exclude such essential items.

18. Section 128.6 (c). This subsection requires that non-product contact surfaces of equipment shall be "kept free of accumulation of dust, dirt, food particles, and other debris." This provision, since it applies to non-product contact surfaces, should be qualified to read "shall be kept reasonably free of accumulation of dust, dirt, food particles, and other debris."

Also, there is a provision which requires "suitable facilities for cleaning shall be provided at convenient locations." Such a requirement would not be necessary as long as equipment is properly maintained. This requirement should be deleted.

19 Section 128.7 (a). This subsection provides that "water used for washing, rinsing or conveying of food products shall be of potable

quality and shall not be recirculated unless suitably treated to assure its potability." This requirements is not necessary to GMP. In some locations, besides being uneconomical, it could actually result in a water shortage. Nor is it necessary for good manufacturing practice that every operation use potable water. For example, in vegetable plants, a counter current washing system is used. Water used in a final rinse of vegetables moves forward in the processing operation and is used for gross cleaning of field dirt and other matter. In each case, the final rinse for such vegetables is potable water. This section could be satisfactorily corrected to read as follows: "water used for <u>final</u> washing, rinsing, or conveying of food products shall be of potable quality. Recirculated water will be permitted for gross cleaning of products provided such use is not likely to contaminate the finished product."

20. <u>Section 128.7 (b)</u>. This subsection requires that containers and carriers of raw ingredients shall be inspected on receipt. This requirement is not practical under all circumstances; such as, receipt of a carload of corn or grain. This requirement should be qualified by the words "where reasonable and practicable."

21. <u>Section 128.7 (e)</u>. Why is this subsection necessary? The provisions of this subsection seem to have been fully and adequately covered by Section 128.6 (d).

22. <u>Section 128.7 (g)</u>. This subsection requires that all foods and ingredients that may have become contaminated shall be "rejected." If the word "rejected" is used in the sense of destruction, it is objectionable. Many products which do not meet quality control requirements can be reprocessed to bring them into compliance.

23. <u>Section 128.7 (i)</u>. This subsection relates to record keeping and product coding. As written in these proposed guidelines, this section is completely unclear as to what is expected or required. If it relates merely to date coding of the finished product package, this provision would not be objectionable. However, if it anticipates coding and record keeping which could trace specific lots of raw materials to specific batches of product, and specific batches of product from the point of manufacture to the ultimate customer, it would result in an almost impossible situation. This provision must be clarified.

Also, the requirement of records retention for a two year period is unnecessarily long. Most food products are consumed well within a year of manufacture.

24. Section 128.7 (j). This subsection requires that storage and transport of finished products shall be under such conditions as will "preclude all" contamination. Probably no conceivable condition, short of sterile canning of food, could preclude all conceivable contamination. The language of this subsection should be changed to read "storage and transport of finished products shall be under such conditions as will minimize exposure to contamination***."

25. Section 128.8 (a). This subjection again imposes an absolute condition which would, in some cases, be extremely difficult to enforce. It is certainly appropriate for a company to set up a program which attempts to control the factors covered by this provision. But it should not require, in effect, a physical inspection of each employee each day. A provision to the effect that persons shall not knowingly be permitted to work with communicable diseases, boils, sores, etc. would be satisfactory.

26. Section 128.8 (b) (2). The only objection here is to any requirement for the removal of plain wedding bands. Many employees either would not or could not remove such bands and there would seem to be little need for such removal.

Respectfully submitted,
GENERAL FOODS CORPORATION

By *Murray D. Sayer*
Murray D. Sayer, Attorney

Example: Technical Business Letter (Block Form, Open Puncutation)*

INDEPENDENCE MALL WEST, PHILADELPHIA, PA. 19105   TELEPHONE (215) 592-3000

ROHM AND HAAS COMPANY

April 22, 19___

Mr. Harold Thompson,
ABC Manufacturing Company
P. O. Box 655
Greensboro, N.C. 27410

Dear Mr. Thompson:

We have received your letter of April 14, 19___. In response to this inquiry regarding foam coatings, I will outline the technology involved in this application in this letter.

An emulsion which we supply which has found application in crushed acrylic foam coatings for various fabrics is Rhoplex E-358. It offers durability to multiple washings and drycleanings and, when formulated properly, will produce a relatively tack-free coating.

Formulations for preparing a suitable crushed acrylic foam coating vary considerably and are especially dependent upon differences in mill equipment. Other considerations are level of opacity, low temperature flexibility, softness, and surface tack.

A typical crushed acrylic foam backing formulation would be as follows:

Product	Percent (as supplied)
Rhoplex E-358[1]	100
Filler (clay, talc, aluminum hydrate)	20
Titanium Dioxide - Rutile (50% predispersed)	6-10
Ammonium Stearate (33%)[2]	10
Aerotex MW[3] (optional)	2-4

---

[1] Rohm and Haas Company
[2] Nopco Chemical Company
[3] American Cyanamid Company

---

*Reproduced with permission.

The above formulation should be compounded in the order listed. Special emphasis should be placed on completely dispersing the fillers by using a good dispersing-high speed mixer. Thickeners should be employed to raise viscosity to approximately 1200 cps. to minimize filler settling and to achieve the optimum in foam coating properties. The pH of the mix should be adjusted up to 8.5 with ammonium hydroxide. The Aerotex MW melamine resin can be added to further reduce surface tack.

The compound mix is then ready to be foamed by means of an Oakes Foamer or a Firestone Foamer. The mix is pumped into the foaming device where it is aerated and reduced in density to 0.13-0.16 grams per cc. This foam is then doctored onto the fabric to be used at a coating station employing a perpendicular knife. The foam is applied at a wet film thickness of approximately 55 mils.

The foam coated fabric is then picked up by a tenter frame and passes into a drying range where the moisture is removed. Drying temperature and production speed will vary with the length of the dryer and the casting thickness of the foam. However, at a thickness of approximately 55 mils, the coated fabric would require a 250-275°F. temperature for about 60 seconds.

After moisture has been removed from the foam coated fabric, the goods pass through a large crushing mechanism where the cellular nature of the foam is destroyed. Between 50 and 100 pounds per linear inch is usually adequate. If the foam is not adequately dried by the time it reaches the pressure rollers, the foam will pick off fabric. Furthermore, if the foam has been overdried (actually cured to some degree) the crushing will be largely ineffective since the partially cured foam will have a tendency to spring back.

It is extremely important to properly adjust the drying of the foam so that these production problems will be avoided.

After the foam coating has been crushed by the pressure rolls, it then passes into a curing oven at 260-300°F. for approximately 3 minutes. This will crosslink the emulsion in the crushed state allowing it to maintain this appearance as a finished product.

The attached design of a typical crushed acrylic foam coated fabric production unit should better help you understand the present state of the art that I have briefly discussed in this letter.

If we can be of any further assistance to you, please do not hesitate to contact either our technical representative in your area, Mr. C. F. Michaud, or this office.

>Very truly yours,
>
>Joseph E. Young
>Textile and Paper Chemicals Department

JEY:dh

Mr. C. F. Michaud
Rohm and Haas Company
1238 N.W. Glisan Street
Portland, Oregon 97209

Chapter 12
THE MEMORANDUM

The memorandum is a brief report covering a single subject. Memoranda are usually used within a company for internal communication, but many firms adopt the memo form in providing, for example, up-to-date product information to their regular customers on a continuing basis. The memorandum used internally is the accepted form for committing to the written record and file a piece of information of permanent or semi-permanent usefulness in connection with a particular project or administrative procedure of either a routine or special character.

Memos vary in length from the very brief order of several sentences or paragraphs to four or five pages and longer. Since the purpose is chiefly for filing and record keeping, a memorandum is limited to cover just one subject. One thinks of the memo as a helpful form of written reminder of facts that have been established or decisions that have been made, which affect or are of interest to persons within or outside of a business or professional organization. Subjects covered generally require periodic or progress reporting or are of an administrative or directional nature. A memo will announce its information clearly and logically with specific details and will set forth explicitly areas of concern and responsibility.

While the heading of the memorandum is fixedly conventional, considerable variation may occur in the form of the body presentation. As with the letterhead and stationery of his particular company, the writer is provided with a printed sheet containing the "To" and "From" words on it, along with the word "Subject" and frequently an imprinted blank for the date and file or reference number. These elements call for no more than the filling in of blanks.

The contents of a memorandum are arranged and presented according to the nature and extent of the subject, the practices of the company, and the preferences of the writer. Paragraphs may be blocked or indented, although the block arrangement is preferred and more common. Headings, subheadings, numbering, lettering, and outlining may be used. The general principles relating to conciseness, clarity, style, and tone of all good writing, technical and nontechnical, are adhered to.

Memoranda are usually not signed by the sender, yet many writers prefer to initial the sheet in ink, either near their typed-in name at the "From" heading or at the end. Conventional practices regarding the initials of the writer and typist, the listing of enclosures, and the names of persons to receive copies are followed. As with the standard business letter, this information appears at the left margin, several spaces after the final line of the message.

Example: Technical Memorandum

INTER-OFFICE LETTER

To: F.D.L.
Place: New York

DATE: March 30, 19
FROM: W.P.B.
PLACE: Atlanta

SUBJECT: Trip Report: American Crystallographic Association Meeting, New Orleans, Louisiana, March 2-5, 19

The winter meeting of the ACA was held in the Chemistry Department of Tulane University and therefore it was fitting that the associated symposium was concerned with experimental and theoretical studies of intermolecular and interionic forces. These forces were discussed in terms of thermochemistry and crystal energies, vibrational spectroscopy and lattice dynamics. In addition, the use of potential functions and packing theory were considered in the solution or prediction of crystal structures, an area in which considerably more work is required. Mathematical techniques were in general emphasized, including density matrix theory and multiconfiguration self-consistent field theory. Discussion also included the effects of electromagnetic fields and spin terms in the Hamiltonian (with applications in ESR and NMR).

Instrumentation of interest to the Laboratory included an X-ray image detector and intensifier which P____ Corporation now offers for sale. This video display system allows instantaneous viewing of X-ray images with fields of view of 4-8 cm diameter by coupling the response of an X-ray sensitive phosphor by means of fiber-optics to an image-intensifier tube and a secondary electron conduction vidicon tube. The signal from the latter is then displayed on a television monitor. Sensitivity is such that diffraction spots produced by 5 CuK alpha photons per frame can be observed at standard TV framing rates. Resolution of the system is presently limited by the resolution of the X-ray phosphor to about 30μ. This type of instrument is receiving considerable attention in various laboratories as evidenced by S.L.B.'s report on the 12th Scintillation and Semiconductor Counter Symposium and by our own request of last year to pursue work along these lines. One company at present offers a simplified but severely limited display system with no fibre optics for $3500. It cannot be over emphasized that this type of instrumentation will find wide application in laboratories as well as in in-line quality control of mill processes such as tube drawing.

P____ Laboratories have interfaced a Norelco powder diffractometer modified with a Geneva-gear stepping device with a small computer to form an automatic system from which d-spacings and relative intensities are obtained directly. This eliminates the manual reduction of data as required by conventional diffractometers.

The International Union of Crystallography is initiating its third experimental study. This will deal with the changes in integrated intensities which have been observed for certain single crystals with continued exposure to X-rays. This variation of intensity with X-radiation exposure has become particularly apparent with the widespread adoption of automatic single-crystal X-ray diffractometers.

W P B
W. P. B.

WPB:cal

Example: Technical Memorandum (Response to Customer Inquiry)*

250 NORTH STREET, WHITE PLAINS, N. Y. 10602

CALUMET BAKING POWDER INFORMATION*

Thank you for your recent letter requesting information for your school project on baking powder.

If you would like to make a display, one thing you could do would be to show the various ingredients which are used in baking powders. A description of the three types of baking powder and how they differ follows. All of the ingredients used in the various baking powders are listed on the package label. Most, if not all, of them can probably be purchased in your local drug store, or they may be among the supplies which your teacher has. If neither your teacher nor the drug store has them, perhaps your teacher could tell you where they could be secured locally.

For demonstration purposes, you might like to make up several cakes, using baking powders you can make yourself as well as Calumet. (A cake recipe for you to follow is attached.) To do this - -

(a) make a baking powder composed of 37% monocalcium phosphate, 30% soda, and 33% starch. The cake you make using this baking powder will be over-leavened because of an incorrect balance of ingredients which gives too much gas (CO^2). As a result, the cake will dip in the center and be coarse-grained. This baking powder, in addition to causing too much gas, develops it too soon.

(b) make a baking powder composed of 33% sodium aluminum sulphate, 30% soda, and 37% starch. The cake you make using this baking powder will be under-leavened because too little gas is produced too late. It will also lack volume. This baking powder will have too slow an action for the baking time.

(c) use Calumet Baking Powder according to the recipe directions. We believe you will be pleased with the result!

The chemical reactions for the above baking powder mixtures are approximated by the following:

---

*Reproduced with permission.

(a) $3\ CaH_4(PO_4)_2 + 8\ NaHCO_3 = CA_3(PO_4)_2 + NaHPO_4$
    $+ 8\ CO_2 + 8\ H_2O$ (25.6%)
(b) $2\ NaAl(SO_4) + 6\ NaHCO_3 = 2A1(OH)_3 + 4\ NA_2SO_4$
    $+ 6\ CO_2$ (26.7%)
(c) This reaction is, in essence, the summation of both a + b.

We hope the above information is helpful to you in your project.

Example: Technical Memorandum

DU PONT
ELECTRONIC PRODUCTS

E. I. DU PONT DE NEMOURS & CO. (INC.) • ELECTROCHEMICALS DEPARTMENT • WILMINGTON, DE 19898

COMMERCIAL PRODUCTS INFORMATION

CERTI-FIRED* THICK FILM RESISTOR COMPOSITIONS
7800 SERIES *

The 7800 Series CERTI-FIRED Thick Film Resistor Compositions are based on the palladium-palladium oxide-silver-glass system. The outstanding features of the 7800 Series include:

- Low cost
- CERTI-FIRED specifications assuring lot-to-lot reproducibility of electrical properties
- Relative insensitivity to firing temperature and time
- A long history of reliability in use

Important properties of the 7800 Series Resistor Compositions include:

- Working resistivity range of 10 to 100K ohms/square
- Typical TCR's of <300 ppm/°C
- Typical drift of less than 1% after 1000 hours of loaded or high-temperature storage

These features and properties make the 7800 Series Resistor Compositions useful in a great variety of thick film microcircuits. The 7800 Series includes ten CERTI-FIRED thick film resistor compositions. New compositions to extend the series and improve the processing characteristics are under development. Please consult your Du Pont representative for current information.

* Du Pont trademark
A-66898   Printed in U. S. A.

THE INFORMATION GIVEN HEREIN IS BASED ON DATA BELIEVED TO BE RELIABLE, BUT THE DU PONT COMPANY MAKES NO WARRANTIES EXPRESS OR IMPLIED AS TO ITS ACCURACY AND ASSUMES NO LIABILITY ARISING OUT OF ITS USE BY OTHERS. THIS PUBLICATION IS NOT TO BE TAKEN AS A LICENSE TO OPERATE UNDER OR RECOMMENDATION TO INFRINGE ANY PATENTS.

---

*Reproduced with permission of the Du Pont Company.

Technical Memorandum (Cont.)

CERTI-FIRED* Thick Film Resistor Composition	CERTI-FIRED Specifications	
	Resistivity ($\Omega/\Box$, ±30%)	TCR, 25°–125°C (ppm/°C, max.)
7800 (blending composition)	<1	---
7810 (blending composition)	1-10	+400
7822	40	+300
7823	640	+300
7828	3K	+400
7837	12K	+200
7889	100K	+250
7826	250	+650
7827	1.5K	+500
7832	7K	+300

Resistor compositions 7826, 7827, and 7832 have been superseded by the improved compositions 7822, 7823, and 7837.

The 7800 Series Resistor Compositions are formulated for application by screen printing. Under identical conditions, all compositions give about the same print thickness. Firing in air to a peak temperature of about 760°C generally gives the best combination of fired properties. Intermediate resistivities are obtained by blending 7800 Series Resistor Compositions. All 7800 Series Resistor Compositions are generally compatible with commonly used Du Pont platinum-gold, palladium-gold, or palladium-silver conductor compositions.

Composition Properties

All 7800 Series Resistor Compositions meet the following specifications:

Total Solids	65.0-68.0% by weight
Viscosity	170M-230M cps (Brookfield RVT, Spindle #7, 10 rpm, 25° ±1°C)

The recommended thinner for these compositions is Du Pont Electronic Composition Thinner 8250. Thinning 7800 Series Resistor Compositions is not recommended, but may be necessary periodically to replace lost solvent.

Recommended Application and Firing Procedure

In order to manufacture reproducibly thick film resistors of acceptable quality, a certain degree of control of the manufacturing processes is necessary. The detailed procedure below describes the conditions under which optimum performance of the 7800 Series CERTI-FIRED* Thick Film Resistor Compositions has been obtained as well as the degree of control necessary for critical processing parameters.

Suitable termination materials for 7800 Series Resistor Compositions include Platinum-Gold Conductor Composition 7553, Palladium-Gold Conductor Compositions 8227 and DP-8267 and Palladium-Silver Conductor Composition 8151.* The termination materials are screen printed first and separately fired to the optimum temperature for the conductor composition involved. Co-firing with some palladium-silver conductor compositions is possible, but not recommended for best results.

The resistor composition is screen printed over the termination with a 165-mesh stainless steel screen. The recommended dried print thickness for all 7800 Series Resistor Compositions is 30 microns which can generally be obtained under these printing conditions. Dried print thickness should be controlled to ± 3 microns to ensure reproducible fired properties.

Prints are dried 5-15 minutes at room temperature to permit leveling of the screen mesh pattern, then thoroughly dried at 125°-150°C with radiant or convection heaters.

The dried prints are fired through a conveyorized furnace. A 45-minute cycle with a 6-7 minute peak at 760°C is recommended. The optimum firing schedule, particularly rise rate and peak temperature, may vary with the particular type of furnace used. Peak firing temperatures of 725°-780°C can be used, but best results are generally obtained between 730° and 760°C. The furnace used should be capable of accurately maintaining the selected temperature profile. Peak temperature should be regulated to at least ± 2°C.

Because of the chemical reactions which occur during firing, it is also necessary to control the firing atmosphere. A sufficient, steady flow of clean, uncontaminated air through the furnace ensures that there will be oxygen available to burn off the organic matter during the heating

---

*Consult your Du Pont Electronic Products Representative for information on compatibility with more recently developed conductor compositions.

portion of the cycle and react to form the desired resistive materials during the peak temperature portion of the cycle. (Particularly detrimental contaminates include chlorinated hydrocarbons which are found in many commercial solvents.)

If the fired resistor is to receive rough handling or be exposed to severe atmospheres during subsequent processing or in use, glass encapsulation of the resistor is suggested. Du Pont Glass Resistor Encapsulant Composition 8185 is recommended for this use. Encapsulation with 8185 produces predictable changes in resistance of 3% or less and protects the resistor from abrasion, humid or mildly reducing atmospheres, reactive organic fluxes, encapsulants, potting compounds, and overspray from air abrasive trimming.

Typical Fired Resistor Properties

Typical fired resistor properties are based on laboratory tests using the recommended application and firing procedure. Unless indicated otherwise, the following conditions were used in preparing samples:

Substrate	96% Alumina
Terminations	Platinum-Gold Conductor Composition 7553 prefired at 1000°C
Printing Screens	165-mesh stainless steel
Firing Schedule	45-minute cycle, 760°C peak for 6-7 minutes (see Figure 1)
Resistor Pattern	2☐'s, 100 x 200 mils

Attached is a tabulation of the more important electrical properties of the 7800 Series CERTI-FIRED* Thick Film Resistor Compositions. Also attached are graphs showing typical firing profile, blending curves, resistivity as a function of dried print thickness, and changes in resistivity with temperature.

Complete details of processing 7800 Series Resistor Compositions are contained in the *Thick Film Microcircuitry Handbook*. Also included in this reference are complete results of many studies made of electrical properties and how they are influenced by variations in processing conditions. Your Du Pont representative should be consulted for additional information or assistance in obtaining a copy of the *Thick Film Microcircuitry Handbook*.

---

*Du Pont trademark

Chapter 13
THE TECHNICAL ARTICLE

Current ideas relating to new and established technologies are recorded, exchanged, and made known through the technical article. Each field of science and technology has professional societies and organizations which sponsor the presentation of papers at meetings. There are countless professional journals and periodicals through which technologists obtain current information relating to their special interests. There is even a larger number of semi-professional popular commercial magazines read by general readers whose personal business interests are in some way related to various phases of technology. The technical article is the main instrument for presenting information that informs experts and nonexperts about current developments and thinking in technological areas.

Since communication and exchange of information is central to technology, articles strive to report fresh findings or to report advances in the application of knowledge. Writers of technical articles offer new slants on established ideas, make fresh commentaries and interpretations of existing information, and freely offer individual points of view which are so necessary to the general growth of useful information.

There is not a great difference between a standard technical report and a standard technical article except in degree. If a report writer has pressure on him to be clear, accurate, and concise in his writing, the article writer has even more so. The article, in a manner of speaking, is public property. It places the writer before not only his professional peers but the public in general. Errors in fact and deficiencies in writing or presentation carry a proportionately greater penalty.

GUIDELINES TO ARTICLE WRITING

Each journal or magazine for which a particular article is intended has its individual editorial requirements and standards. While these are sometimes available to the writer from the journal in printed form, more often the writer must determine them by careful study. It is simple enough to determine whether an abstract accompanies each article, or what form and position bibliographic citations take in a particular journal or magazine. There may even be an established format of order for the presentation of all articles. The kind and number of graphic illustrations–charts, graphs, formulas–may similarly be assessed by concentrated study of the journal. A journal editor may be willing to make minor alterations in an otherwise acceptable manuscript, but the technical writer who does not rigorously tailor his article to the existing practices of the publication will meet with categoric rejection.

Each editor and, therefore, each writer must ask himself the following questions: Is the subject valuable enough to warrant publication? Does the writer know his facts thoroughly? What kind of graphic illustrations are required? What form is followed in presenting footnotes and bibliography? Each article must be thought out in detail and presented impartially and objectively. With respect to style, the writer must strive to produce sentences, paragraphs, diction, and an overall design of outline in keeping with established publication standards of the journal or magazine in question. Editors demand this, readers expect this, and writers must accept this. An article must be shaped to the requirements of the journal and to the needs of its audience.

Emphasis is upon the essential. Printing and editorial costs today are too high to allow reproduction of nonessential, unclear, inaccurate, or ambiguous information. Definitions must be direct and analyses must be trimmed to logic. The writer may very well know his subject in broad outline, but is his understanding deep enough to explain it in clear detail for a reader? The value of writing as a severe but invaluable discipline in the thinking process becomes evident to the technician writing an article for publication.

Example: Short Technical Report (A Magazine Article)

Servicing Cassette Recorders* *By Homer L. Davidson*

Today's cassette tape recorders are produced for mono and stereo track cartridges. The deluxe recorder will tape in four track stereo and mono positions. Small portable units record and play two track mono on easy-loading, drop-in cartridges. Many of these units will record up to two hours at 1 7/8 ips, on both tracks.

The typical portable cassette may be only a player or a combination recorder and player. Most units have a frequency response from 80 Hz to 10 kHz. Some units are battery operated or operate from ac power line.

The new cassette tape units have many features: level meter, input for radio and output for external amplifier, automatic level control to reduce distortion caused by overloading, convenient "pop-up" cassette loading, remote control mike or built-in mike and speaker, and some deluxe cassette home systems have push-button controls with a digital counter for indexing. Still others may have satellite speakers and can be used as a PA system.

Circuit Description

The portable cassette recorder or player amplifier may have from five to ten transistors and only one tape head while a larger stereo recorder may have up to thirty transistors and diodes. Most cassette recorder circuits are comparable to the reel tape recorder.

The unit shown in Fig. 2 has a separate erase head and oscillator stage. In the playback position, the record/play head is switched to the input circuit of a pre-amplifier stage. Three preamplifier stages are used in this particular model with a volume control located between second and third pre-amp stages.

The audio driver, Q4, is directly coupled to the third pre-amp circuit with the driver output signal transformer coupled to a push-pull output circuit. An eight ohm PM speaker is switched into the voice coil circuit.

---

*Reprinted from the February 1970 issue of *Electronic Technician/Dealer* by special permission from the copyright owners, Harbrace Publications, Inc. (Graphics and figures accompanying the original article have been omitted.)

The portable cassette recorder may be either battery operated or ac operated. Most ac power supply circuits employ a step-down power transformer and silicon diode fullwave rectification (Fig. 2). Large electrolytic capacitors and choke or resistance filtering networks are also found in the ac power supply circuits. Capacitor C27, and choke L2 are the filter components (Fig. 2).

Speed Problems

If the speed of a player varies, always re-check player operation with a new cartridge. Listen to the pitch of the recording to see if the cassette player is running fast or slow. In most cases the tape is running too slow.

Since it is possible to have a defective cartridge, substitution is the fastest test. Also, when some cartridges are played for a number of hours, the small tape roller inside the cartridge may become sluggish. A drop of oil on each side of the rubber roller will cure the slow or erratic tape cartridge.

Tape speed problems can also be caused by a defective belt and capstan pulley assembly. Check the drive belt for stretching or for oil spots. Clean the belt and drive assembly with alcohol or a non-toxic cleaner. See if the belt is properly aligned with the capstan drive pulley. In some cases the capstan drive assembly mounting screws may become loose and move the pulley out of proper alignment. A cracked or broken belt should be replaced with an exact part. In some of the small players the motor drive belt looks like a small rubber band (Fig. 3). We don't suggest a rubber band as a belt replacement.

Cleaning and Lubrication

Tape heads should be cleaned periodically to keep them in tip-top shape. Anytime a cassette recorder comes in for repair, a complete clean up job should be made. Clean the tape head and guide assemblies with alcohol and a "Q" tip. A can of tape head spray will remove most tape dust residue. Also clean any tape oxide from under the pulleys, flywheel and tape heads. Remove the motor and pulley drive belts and wipe them with alcohol and a clean cloth. (A special tape head cleaning cartridge is also available which is simply run through the player like an ordinary cartridge.)

Remove the capstan flywheel and clean the top bearing. A "Q" tip or round brush are useful for removing oxide dust from inside the capstan bearing. Wipe the bottom nylon bearing clean and apply a coat of lubriplate grease. A spot of oil on small motor and pulley bearings is sufficient. Small motors found in portable cassette recorders are normally self-contained and do not require lubrication.

Complete the cleaning operation by spraying the operation switches and relay points. A thin piece of cardboard will also help to clean up those dirty relay points, but do not use an abrasive material such as sandpaper on the contacts.

Checking the Amplifier

A cassette recorder that will not record or playback is possibly the victim of a defective tape head or amplifier. First, try a new recording and make sure the cartridge is properly seated. If the recorder will playback a new cartridge but will not record, suspect a defective microphone and cable. You can assume the play/record tape head is functioning.

Check the amplifier by injecting a signal from an audio generator at the tape head terminals. Signal trace the defective amplifier by going from the base to collector terminal of each transistor and make voltage and transistor tests of the circuit (Fig. 4) where signal is lost. When signal tracing from stage to stage notice the loss or gain of each stage. Compare the injected signal at each end of a suspected electrolytic capacitor. Remove it and take a leakage reading. A good electrolytic capacitor will cause the meter hand to charge upward and slowly decrease. Leaky capacitors will show a lower resistance reading and should be discarded (Fig. 5).

Always signal trace the audio circuit by starting at the output stage and work toward the input when the amplifier appears intermittent. Start at the collector circuit and then apply the signal to the base terminal. If the signal "pops on and off," a transistor is intermittent. A freeze can be used on a suspected transistor to see if it becomes intermittent. Several applications of freeze spray may be needed to obtain the intermittent condition.

A defective transistor may also be located with voltage measurements. When the collector voltage is higher than normal, suspect a transistor that is not conducting or has open emitter bias resistor. With all voltages close to the same potential suspect a leaky transistor. A low collector voltage also can indicate a leaky transistor or improper bias. When the collector voltage is the same as the supply voltage the transistor is open. Use a meter with a high input impedance to make voltage tests.

Another way to check a suspected transistor is with an incircuit transistor tester. Remember that an open transistor will not indicate a beta reading and in directly-coupled circuits, it is best to remove the collector lead for more accurate leakage tests. If in doubt, remove the transistor from the circuit and make your tests. The record/play tape head can be checked for opens or shorts by taking a resistance reading. These readings will vary from 50 to 500Ω and should be checked against the schematic. When in doubt substitute a new head. In stereo units, a defective head in one channel can be checked by swapping heads.

Hum and Distortion

Hum appearing on a portion of a cartridge that has normally recorded material ahead of it can be caused by a bad mike cable or connection. Check the cable for possible breaks at the plug or close to the microphone case. Flex the mike cable while making resistance measurements from plug to mike connections. Excessive hum can also be produced by a worn spot in the volume control, a poor ground terminal or a broken foil on the ground section of a circuit board.

Distortion conditions are often caused by leaky output transistors or improper bias resistors. "Crosstalk" distortion can be removed with proper height adjustments upon the tape head. Excessive recording volume, with excessive bass response, will also produce a distorted recording as will a tape head caked with oxide dust. Excessive tape noise may be caused by a magnetized tape head.

Check the audio output transistors for possible distortion by taking accurate voltage readings across the base and emitter bias resistors. Be sure to replace any bias resistors with the exact values. Power output transistors should be removed from the circuit to make a beta leakage test. Often it will save valuable service time to replace both output transistors. Make sure the base and emitter resistors are the correct value before replacing the power transistors (Fig. 6).

Leaky directly coupled preamplifiers or interstage transistors are also candidates for distortion. Make careful voltage measurements at the transistor and remove the collector terminal to test with in-circuit beta transistor tester. Make in-circuit beta leakage tests on both output transistors in a push-pull stage.

Head Adjustments

Excessive crosstalk and poor frequency response can result from improper tape head alignment as previously indicated. Height adjustment screws are usually located on each side and to the rear of the record/play tape head. Many of the small cassette tape player heads are stationary and adjustment requires that the tape head be bent or twisted into position. Some of the low priced cassette players have a piece of rubber underneath the tape head and one side screw that will allow for the azimuth adjustment.

In the higher priced models side tension springs with adjustment screws are provided for both height and azimuth adjustments. The height adjustment is made to eliminate crosstalk while the azimuth adjustment is set for the best frequency response. The tape head adjustment screws in higher priced cassette units are similar to those used in regular tape recorders. The following is a handy list of checks you can make to insure the overall operation of the cassette unit.

1. Completely demagnetize the tape head after servicing the unit.
2. Automatically clean the tape head and guide assemblies with a "Q" tip and alcohol or audiotex blast off head cleaner.
3. Give the cassette recorder a complete operation test with test tape or a known recording. Check out each phase of operation.
4. Clean up the entire cabinet and plastic parts with foaming window spray. A convenient troubleshooting chart is provided on the next page.

Example: Short Technical Report (A Magazine Article)

**House Painting
with a Roller***

Painting the outside of your house is a dreary job that most people want to complete as quickly as possible. A job that normally would take three days can be completed in little more than one if you use a roller instead of a brush.

No matter what type of siding you have on your house, a roller will lay down a good covering of new paint. However, one of the most important factors in getting a satisfactory job with a roller is to select the correct tool and to make sure it is of the highest quality.

Use a 9-in. roller for all types of siding except clapboard. Since clapboard is relatively narrow, you will have to use a 4- or 5-in. roller, depending on the siding's width.

Proper nap length is important to assure complete coverage. For smooth surfaces such as clapboard and smooth shake siding, a 3/4-in. nap should be used. If the surface is rough wood, such as serrated cedar shakes, use a 1-in. nap. A 1 1/4-in. nap provides the best coverage for rough, porous surfaces such as stucco, brick and other masonry.

The nap material is a primary consideration. If you are using water-base paint, buy a roller of Dynel and nylon blend. These synthetic fibers won't absorb the paint. A roller made of wool or cotton will be

---

*From *Mechanix Illustrated*, April, 1968. Reprinted with permission from *Mechanix Illustrated* Magazine. Copyright 1968 by Fawcett Publications, Inc.

ruined if used in latex paint and will not do a good job. For oil-base paint, use either a Dynel-nylon blend or pure lambskin shearling.

A bird-cage roller frame is best. Its wire construction is easier to clean than a solid frame and the roller sleeve is less apt to freeze with drying paint.

The price of a roller is a good indication of its quality. Use the best for this big job. A top-flight 1 1/4-in. Dynel-nylon roller costs $2.50 to $3. A good-quality 1 1/4-in. lambskin roller sells for $3.75 to $4. Prices are slightly less for shorter naps. A well-constructed bird-cage frame costs about $2.

Preparing the surface for painting is no different with a roller than with a brush. However, brush bristles disturb the old chalky surface and lay the new paint against the substrate for good adhesion. A roller won't do this. You should, therefore, make sure the surface is as chalk-free as possible. Sand the siding and wash it clean with detergent. Rinse thoroughly with water.

The type of stroke you use when rolling depends on the type of siding on the house. Use a horizontal roll to paint smooth siding—clapboard, plain shingles, unserrated shake, plank, etc.

With vertical planks and all serrated siding, such as cedar shake and asbestos shingle, you must roll with the serrations to make sure paint is applied between them. Thus, you roll vertically. Rolling paint onto stucco and brick is done in any manner that's comfortable.

When painting, make certain the roller is completely wetted but not dripping. A roller that is dry in spots will produce holidays; that is, unpainted areas.

Dip the roller frequently but be sure to roll out excess paint. The best way is with a grid, available in paint stores. Place this perforated metal screen in the top of the paint tray. Dip the roller in the paint well and then roll it over the grid to remove the excess.

One thing to watch when using a roller is spatter. You will get more than when using a brush. To minimize spatter, don't roll too fast. When working around shrubbery and walks, cover them with drop cloths.

A long-nap roller makes short work of painting beneath the lips of clapboard and cedar shakes. The nap extends into them, covering the area with paint as you roll. However, keep a small trim brush or conical roller handy to spread paint out across the lip, should it start to drip.

Extension poles 1 to 3 ft. long allow you to extend your reach when on a ladder. Poles up to 8 ft. may let you get away without a ladder but don't use longer poles. They are tough to handle and create excessive spatter. Extensions screw onto the roller spindle once the handle is removed.—*Bud Weis.*

Example: Technical Article For Semi-Technical Audience* (Bibliography Included)

High Radium-226 Concentrations in Public Water Supplies*

To identify population groups consuming water with a relatively high radium content (more than 3.0 picocuries per liter), selected public water supplies were analyzed for radium-226. (The term "radium" indicates radium-226 unless otherwise specified.) The principal objective of the study was to identify compact populations which could serve as a base for epidemiologic studies of the effects of chronic radium ingestion at low levels.

Marinelli (1) pointed out earlier the possibilities of using such populations to study the effects of radium ingestion. One such study by Petersen and co-workers (2) attempted to relate the mortality rates from bone neoplasms to the level of radium in drinking water for 111 midwestern cities. These authors reported that a retrospective analysis of death certificates indicated that the mortality rate from bone neoplasms in a population ingesting an average of 4.7 picocuries (pCi) of radium per liter from its water supplies was higher than in a matched control group with almost no radium in its water supplies. Although their study included a population of 900,000, few cases of bone neoplasm were found because incidence of this disease is very low. The small observed differences in the mortality rates between the two groups were therefore not statistically significant. Before a difference of the magnitude found by Petersen and co-workers could be considered significant, many more cases would be needed for study; hence, a larger study population or many more years of observation of the same population would be required.

Petersen and co-workers (2) concluded that the identification of additional population groups exposed to relatively high radium concentrations offered the greatest hope for a more meaningful study. They observed that the feasibility of such a study might be limited by the lack of a sufficiently large exposed population. The National Center for Radiological Health undertook the present study to identify additional population groups who might be exposed to relatively high radium concentrations in their drinking water. Also, the study was viewed as a means of locating water supplies that would require corrective action

---

*From John L.S. Hickey and Samuel D. Campbell, "High Radium—226 Concentrations in Public Water Supplies, *Public Health Reports*, Vol. 83, No. 7, July, 1968. Reproduced with permission.

should epidemiologic studies show that low-level radium ingestion is harmful to human beings.

Selection of Water Supplies

There are more than 19,000 community water supplies in the United States, serving more than 150 million people (*3*). The laboratory facilities available for the present study permitted radium analyses of only approximately 1,000 water samples. It was therefore necessary to select those supplies most likely to contain relatively high amounts of radium.

A literature search was made to locate geographic areas with evidence of relatively high radium levels in water (*4-10*). Smith and co-workers (*6*) reported mean values of up to 160 pCi of radium per liter in some Maine well waters and up to 8.4 pCi per liter in some New Hampshire well waters. Scott (*8*) found radium levels exceeding 3.3 pCi per liter in a large number of fresh water wells in several midwest and Rocky Mountain States and in Kansas and Texas. Lucas (*5*) found that approximately 37,000 persons were consuming water containing more than 10 pCi of radium per liter in 21 communities in Illinois, Iowa, and Missouri. Ground waters in Illinois and Iowa have been sampled extensively by several investigators, who reported elevated radium levels in many municipal well water supplies in these States (*2, 5, 7, 8, 10*).

Analyses of the water supplies (*9*) of the 100 largest cities in the United States, serving 60 million people, showed that only three of these supplies contained more than 1 pCi of radium per liter; the supply of Rockford, Ill., had 2.5 pCi per liter; that of Houston, Tex., 1.3 pCi per liter; and that of Lubbock, Tex., 1.9 pCi per liter. Several State health departments provided data on radium or alpha levels in ground water; the U.S. Geological Survey provided additional data (unpublished report by R. C. Scott entitled "Known and suspected high-radium aquifers in the United States, March 1966).

With the foregoing information as a guide, 20 States were selected in which to sample water supplies for radium content (see map). Illinois, Iowa, and Wisconsin were not included in the study because the well water supplies in these States had already been systematically sampled by others. Because radium levels in well water have been reported to vary over time (*10*), several water supplies (principally in Minnesota) with previously reported high radium or alpha levels were sampled to determine if these levels still existed. Although there was no history of high radium levels in water in Alaska and Hawaii, these States were included on the presumption that some unusual radium levels might be found as a result of such features as volcanic and glacial activity.

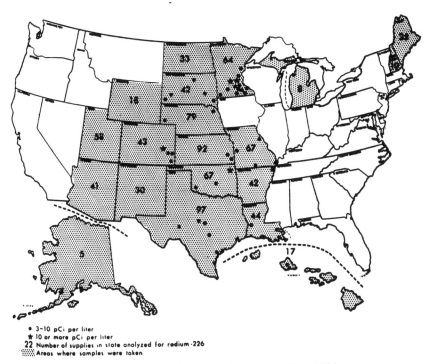

- • 3-10 pCi per liter
- ★ 10 or more pCi per liter
- 22 Number of supplies in state analyzed for radium-226
- ░░ Areas where samples were taken

Radium-226 in public ground water supplies, 1966

Within the defined geographic area, three basic criteria were used in selecting water supplies for radium analysis.

1. Only ground water supplies were selected. Other investigators (4, 5, 7) have shown that surface waters are unlikely to contain significant amounts of naturally occurring radium.

2. Supplies receiving aeration, flocculation, settling, softening, or a combination of these, were not selected. Some types of water treatment have been shown to remove the bulk of the radium (4, 7, 10). Since the principal interest was in water with high radium levels at the point of consumption by the population, treated supplies were not sampled.

3. Supplies serving fewer than 1,000 persons were usually not selected. The 7,850 ground water supplies in this classification serve only 10 percent of the population that consumes ground water (3), and omission of these supplies permitted a broader population coverage with the available analytical facilities. However, 80 water supplies serving fewer than 1,000 persons were sampled, in several areas having particular evidence of elevated radium levels in ground water.

Information was obtained for each community on the source of water, the treatment it received, and the population it served, principally from the 1963 inventory of municipal water facilities (*11*). Both Lucas (*7*) and Scott (*8*) have pointed out that aquifers in certain types of geologic formations are more likely to contain radium than those in other formations. However, there was not sufficient geologic information readily available on specific water supplies to justify using this factor as one of the criteria for selection.

In areas where several water supplies suitable for study were clustered, only one or two were selected in order to obtain broader geographic coverage with the limited laboratory facilities. If any of the supplies selected contained elevated radium levels, it was assumed that the nearby supplies could be more knowledgeably sampled in the future.

Sample Collection

Before the sampling, permission to make the study was obtained from the 20 State health departments involved, through the Public Health Service Regional Offices. State health departments were notified several days in advance when and where samples would be taken. Collapsible plastic containers and return-addressed mailing cartons were mailed directly to the superintendent or manager of each city or community water department. Each envelope contained a request that a water sample be collected and instructions for its collection and mailing. Samples were numbered with the post office ZIP code of the community served.

The requests for water samples were mailed to the 976 communities selected for radium analysis over a period of several weeks in May and June of 1966. Within a few weeks of the requests, 923 samples were received at the laboratory—a response of more than 94 percent. A few of the first samples received did not contain enough water for analysis because the plastic containers had not been expanded sufficiently before filling, and the samples had to be discarded. Brief illustrated instructions were inserted in later requests for samples, thereby eliminating this problem. Of the 923 samples received, 32 could not be analyzed because of insufficient volume or loss of the sample through laboratory error. In one State, the health department selected the communities to be sampled; in two States, the health departments collected and mailed the samples.

Upon completion of the analyses, sample results were sent to each State health department (through the Public Health Service Regional Offices) with a request that the department notify the communities of the results. It was anticipated that questions would arise regarding interpretation of the results. Therefore, concurrently with the study, a digest was prepared of various protection standards and State regulations for radium in water (12). Reprints of this digest accompanied sample results sent to the States.

Results of Sample Analysis

A total of 891 water supplies serving 3,864,500 people were analyzed for radium. Seventy-two water supplies, serving 329,000 people, were found to contain from 1 to 3 pCi of radium per liter. Forty-one water supplies, serving 144,000 people in 10 States, contained 3 or more pCi of radium per liter. Four of the 41 water supplies, serving 3,000 people, contained 10 or more. The highest radium level found was 24.1 pCi per liter in a supply serving 360 people. The remaining 778 samples contained less than 1 pCi of radium per liter of water. These results are summarized in tables 1 and 2. The locations of the water supplies having 3 or more pCi of radium per liter (including the positive confidence interval) are shown in the map. The number of samples analyzed from each State is also shown.

Analytical Methods

Radium analysis was performed under contract during 1966-67 by the New Mexico Department of Public Health. The majority of the samples were radiochemically analyzed for radium-226 using published methods with minor modifications (13). The radium-226 recovery, based on 26 standards of radium-226 (1.0 pCi), was 95±15 percent at the 95 percent confidence level. Early in the study, the laboratory staff had attempted to analyze samples for both radium-226 and radium-228 simultaneously by combining methods described by Petrow (14) and others (13). Analysis for radium-228, however, proved too time consuming for the period allowed for this project, and it was abandoned. The radium-226 recovery with the combined method was 80±20 percent at the 95 percent confidence level.

The mean recovery values found for radium-226 were used in calculating the radium levels reported. Samples for radium-226 analysis were counted for 20 hours. Counting errors were calculated, and the 95 percent confidence limits are reported individually for those samples appearing in table 2.

Discussion

Unpublished information provided by several State health departments and the U.S. Geological Survey was particularly helpful in selecting sampling locations. Two of the four samples showing more

Table 1. Summary of radium-226 analyses of public water supplies, 1966

State	Number of water supplies analyzed	Total population served 1963[1]	Supplies with more than 3 pCi Ra-226 per liter		Highest Ra-226 level observed (pCi per liter)
			Number	Population served 1963[1]	
Alaska	5	9,840	0	0	0.1
Arizona	41	241,780	0	0	2.0
Arkansas	42	178,365	0	0	1.4
Colorado	43	106,750	4	3,110	15.7
Hawaii	17	427,780	1	16,000	8.9
Kansas	92	286,840	2	1,130	5.0
Louisiana	44	299,865	1	9,000	4.0
Maine	25	73,920	0	0	.3
Michigan	8	115,300	0	0	.9
Minnesota	64	304,200	13	46,175	24.1
Missouri	67	175,030	5	17,815	8.7
Nebraska	79	225,820	2	9,660	3.8
New Hampshire	19	59,575	0	0	2.5
New Mexico	30	240,880	0	0	2.3
North Dakota	33	60,740	0	0	.7
Oklahoma	67	202,725	3	6,830	10.3
South Dakota	42	88,930	5	11,550	6.9
Texas	97	539,590	5	23,100	9.7
Utah	58	142,705	0	0	.9
Wyoming	18	83,865	0	0	.8
Total	891	3,864,500	41	144,370	

[1] Source: reference 3.

than 10 pCi of radium per liter (table 2) confirmed results that the Minnesota Department of Health had previously reported for these water supplies. Most of the other water supplies showing 3 or more pCi per liter are located in or near areas with previously reported elevated radium or alpha activity in well waters. Since sampling was generally limited to States having a previous history of elevated radium levels, this result was expected.

Seven widely scattered water supplies in South Dakota and Nebraska showed levels of radium of more than 3 pCi per liter; only two of these, in the southwest corner of South Dakota, were in areas from which elevated radium levels had previously been reported (8). Samples were collected from 44 well water supplies in Maine and New Hampshire; most of these sources were within 50 miles of wells previously reported to have high radium levels (6). However, the highest radium levels found in these 44 water supplies were 0.3 pCi per liter in Maine and 2.5 in New Hampshire.

In southern Missouri, 13 water supplies were found to contain more than 1 pCi of radium per liter; in addition, 15 other supplies in this area had previously been reported by Lucas (5) to contain more than 1 pCi per liter. These results support Scott's postulation (8) that the known high-radium formations of the upper Mississippi valley may extend through southern Missouri. Therefore, this locality may be considered as a promising area for more extensive sampling of ground water supplies for radium.

The next step toward identification of additional populations ingesting high levels of radium from water would be a sampling program for previously unsampled water supplies in the vicinity of the high radium water supplies observed in this study. This step was not possible in this study because of the limited time.

Only those water supplies found to contain 3 or more pCi of radium per liter are named in table 2. This level was used in Petersen's study (2) as the lower limit of radium ingestion from drinking water in defining the "exposed" population. A large number of the water supplies sampled showed virtually zero radium content, and the communities they serve may be suitable control populations in studies of the effects of radium ingestion. Sampling results for all communities included in the study are available from the authors through the National Center for Radiological Health.

The sampling instructions requested that samples be collected on the water distribution systems but not from fire hydrants, deadends, or water taps on which home softeners were in use. This request was made so that the samples would be representative of the water consumed by the population. Because of the previously observed variable levels of radium within a water system (2), several samples from various points on a water supply should be analyzed for radium before a community is selected for an epidemiologic study of the health effects of low-level radium ingestion.

The sample containers and the general method for collecting the samples and getting them to the laboratory proved satisfactory. A large degree of credit should be given to the State health departments for their cooperation in notifying the communities to be sampled in advance and for generally maintaining an overall rapport with their communities, which made the individual water plant managers willing to take the time to collect and mail a sample promptly. This State-city relationship is believed to have been responsible for the high response.

Table 2. High radium-226 concentrations observed in public water supplies, 1966

State and community	Population served, 1963[1]	Ra-226 (pCi per liter)	±95 percent confidence level[2]
Colorado			
Granada	600	3.14	0.07
Holly	1,110	5.65	.06
Walsh	1,000	3.00	.07
Wiley	400	15.71	.15
Wahiawa, Hawaii	16,000	8.90	.08
Kansas			
Mulberry	750	[3]2.98	.05
Walnut	380	4.97	.06
DeRidder, La	9,000	3.95	.08
Minnesota			
Anoka	10,560	7.38	.10
Claremont	465	9.83	.12
Courtland	240	4.82	.09
Faribault	16,925	3.19	.07
Hardwick	330	4.24	.08
Howard Lake	1,005	15.20	.15
Janesville	1,425	7.02	.09
Jordan	1,480	7.12	.11
Montrose	360	24.11	.20
New Market	210	8.83	.12
Red Wing	10,530	4.65	.08
Sandstone	1,550	4.05	.08
Savage	1,095	9.60	.12
Missouri			
Charleston	5,600	4.04	.06
Clarkton	1,440	5.64	.08
Rich Hill	1,555	8.67	.10
Troy	1,650	3.47	.06
West Plains	7,570	4.83	.07
Nebraska			
Kimball	5,160	3.84	.07
Wayne	4,500	3.34	.06
Oklahoma			
Afton	1,250	10.29	.12
Mangum	4,200	7.35	.11
Maud	1,380	5.24	.09
South Dakota			
Canton	2,500	6.93	.11
Edgemont	1,770	5.33	.10
Hot Springs	4,950	4.82	.09
Howard	1,200	3.52	.08
Ipswich	1,130	3.02	.07

Table 2. High radium-226 concentration observed in public water supplies, 1966 (Continued)

State and community	Population served, 1963[1]	Ra-226 (pCi per liter)	±95 percent confidence level[2]
Texas			
Big Lake	2,750	9.71	.12
Brady	5,350	7.72	.10
Eden	1,600	6.21	.10
Kerrville	12,000	3.73	.07
Sea Drift	1,400	3.45	.07

[1] Source: reference 3.
[2] 95 percent confidence level based on counting errors only.
[3] Included because confidence limit range includes 3.0.

Summary

To identify population groups in the United States having relatively high chronic intakes of radium from drinking water, samples of approximately 900 public ground water supplies serving 3.8 million people in 20 States were analyzed in 1966-67.

After the National Center for Radiological Health, Public Health Service, had arranged for the water sampling with the health departments of the 20 States, personnel of these departments or of community water departments collected 1-liter samples of the selected water supplies and mailed them to the laboratory performing the radium analyses. The response to the request for water samples was more than 94 percent.

Forty-one water supplies, serving 144,000 people in 10 States, were found to have 3 or more pCi (picocuries) of radium-226 per liter. Four supplies, serving 3,000 people contained 10 or more pCi per liter. The highest concentration observed was 24.1 pCi per liter in a supply serving 360 people.

REFERENCES

(1) Marinelli, L. D.: Radioactivity and the human skeleton. Amer J Roentgen:80: 729–739 (1958).

(2) Petersen, N. J., Samuels, L. D., Lucas, H. F., and Abrahams, S. P.: An epidemiologic approach to the problem of low-level radium 226 exposure. Public Health Rep 81:805–814, September 1966.

(3) U.S. Public Health Service: Statistical summary of municipal water facilities in the United States, January 1, 1963. PHS Publication No. 1039. U.S. Government Printing Office, Washington, D.C., 1965.

(4) Hursh, J. B.: Radium content of public water supplies. J Amer Water Works Assoc 46: 43-54 (1954).

(5) Lucas, H. F., Jr.: Study of radium-226 content of midwest water supplies. Radiol Health Data 2: 400–401 (1961).

(6) Smith, B. M., Grune, W. N., Higgins, F. B., Jr., and Terrill, J. G., Jr.: National radioactivity in ground water supplies in Maine and New Hampshire. J Amer Water Works Assoc 53: 75–88 (1961).

(7) Lucas, J. F., Jr., and Ilcewicz, F. H.: Natural radium-226 content of Illinois water supplies. J Amer Water Works Assoc 50: 1523–1532 (1958).

(8) Scott, R. C.: Radium in natural waters in the United States. Radioecology. Proceedings of the first national symposium on radioecology, 1961. Reinhold Publishing Corp., New York City, 1963.

(9) Durfor, C. N., and Becker, E.: Public water supplies of the 100 largest cities in the United States, 1962. U.S. Geological Survey Water Supply Paper 1812. U.S. Department of the Interior. U.S. Government Printing Office, Washington, D.C., 1964.

(10) Ball, A. D.: Protocol for proposed plan of study of the causes of variability in radium concentration in municipal water systems. State Hygienic Laboratory, State University of Iowa, Iowa City, September 1964.

(11) U.S. Public Health Service: Municipal Water Facilities, 1963 Inventory. PHS Publication No. 775. U.S. Government Printing Office, Washington, D.C., 1964, vols. 1–9.

(12) Hickey, J. L. S.: Digest of protection standards and State regulations for radioactivity in water. Radiol Health Data Rep 7: 549–554 (1966).

(13) Radionuclide analysis of environmental samples. A laboratory manual of methodology. Robert A. Taft Sanitary Engineering Center Technical Report R59–6. U.S. Public Health Service, Cincinnati, Ohio, 1966, pp. 42–56.

(14) Petrow, H. G., Cover, A., Schiessle, W., and Parsons, E.: Radiochemical determination of radium-228 and thorium-228 in biological and mineral samples. Anal Chem 36: 1600–1603 (1964).

Chapter 14
THE
ABSTRACT

An abstract is an abbreviated statement of a longer piece of writing. The abstract concentrates only on the essential ideas of the longer work, omitting all information used in the original for illustration and development. To present in compact form the central substance of a longer discourse is the aim of an abstract.

Abstracts appear in technology both with and without the reports they summarize. It is the practice in many segments of business and industry to have an abstract accompany all major reports which are prepared for circulation. In such instances the abstract appears at the very head of the report and is usually the first element the reader encounters. A reader is able thus to determine from the abstract whether or not the full report will require reading. Even if the abstract indicates to the reader that the material of the report does not apply to his particular area of responsibility or competency, he will have acquired a passing view of work occurring in related areas of endeavor.

Abstracts appear frequently without the full reports they represent. Certain scholarly publications outside and inside of science and technology devote their whole issues to abstracts of recent work. There are, for example, the journals known as *Dissertation Abstracts* and *Chemical Abstracts*. Readers whose interests are covered by a particular abstract are kept apprised of recent work in their field and may thus consult the original long work as the need arises.

WRITING THE ABSTRACT

The effectiveness of an abstract is determined by the requirements of the reader. He must be able to gain a general overview of the purpose, work, and result of the study. The assumption is that the reader is a busy person who does not have time to read in entirety all of the reports in the field. Yet this same reader is required to keep abreast of developments generally as well as those particularly in his field.

The abstract writer may or may not choose to follow the organization and outline of the original text. It seems plausible that the first sentence or two of any abstract should state cogently and concisely the aim or purpose of the study. This should not require more than one or two sentences. Following the briefly stated purpose may come sentences covering the main work of the study, usually emphasizing the results or findings of the study rather than the processes by which they were reached. The final sentences of an abstract might deal with recommendations and conclusions of the report.

Abstracts are written in the third rather than the first person—"This study seeks to determine"; "Tests were made on specimens"; "Power shortages occur." There is no need for "guidewords" such as "in the view of the author"; "it is the view of the experimenter"; or "Smith maintains." Readers assume that all the ideas advanced in the abstract are those of the study being abstracted—guidewords unnecessarily lengthen and clutter up the picture. Whether the abstracter is dealing with his own or someone else's report, it is always assumed that the ideas and substance of the abstract are drawn directly from the original work, although phrased in the abstracter's own words.

Abstracts which go over 500 or 600 words are rare. While there is no standard rule regarding length, the usual abstract is from 250 to 400 words in length. Brevity and coherence are the main guidelines. The abstracter must recognize and deal compactly with basics and essentials only. The readers' need for a meaningful, panoramic picture must be satisfied.

EXAMPLES: ABSTRACTS*

THE POSSIBLE USE OF A SYSTEMS APPROACH
IN BALANCING INTERMITTENT-FLOW
PRODUCTION LINES.

Leon Wayne Transeau, Ph.D.
The American University, 1968

Statement of the Problem

The problem studied is the possible use of a systems approach in maintaining balance along intermittent-flow production lines. Intermittent-flow production lines, as contrasted to continuous-flow lines, require special attention because inprocess inventories at each operation fluctuate in size and must be held within acceptable limits. Maintaining balance along lines of this type may require the reassignment of workers as workloads along the line change.

Procedures and Methods

Current methods of controlling the balance along intermittent-flow production lines were studied in depth. A generalized system was developed and evaluated with the aid of a computer model designed to simulate production line flow and measure the degree of balance achieved. A case study was made in which the system developed was compared in an actual production situation with the methods in current use.

Results

In simulations made with the model the best overall operating results were obtained with the system developed using proposed control limits. Worker idle time was at the minimum observed level, and production for the line was at the maximum level.

Data collected during the case study indicated that the system tested provided improved results in worker efficiency, production, idle time, and number of reassignments required. Production for the line, which was the critical measure of performance, increased 7 percent with the proposed system.

---

*From *Dissertation Abstracts, The Sciences and Engineering*, Vol. 29, No. 4, October, 1968. Reproduced with permission.

Conclusion

It was concluded that the systems approach would be of practical value in controlling the balance of intermittent-flow production lines under many operating conditions. Although the specific systems developed provided only a beginning point for further research, it was shown that such a system can effectively improve upon the results now obtained in production situations where trial-and-error techniques are currently used.

THE RELATIONSHIP BETWEEN THE MACHINE-TOOL STRUCTURE AND ITS COST.

Isaac Moked, Ph.D.
University of Illinois, 1967

The complexity of modern machine tools introduces severe decision making problems, not only for the buyers and users of machines, but primarily for their producers.

These problems, involving evaluation of design and economic alternatives, often require precise solutions in a very limited time.

The frame of a scientific method for the evaluation of design and economic merits of machine tools and their elements based on a systems analysis approach, is proposed. It involves the systematization of the analysis, forecasting and decision making functions of a machine tool company; it provides the policy formulating system with accurate, methodically obtained and correlated information on optimal structures of machine tools to meet the consumers' demand and maximize profits.

The problem is formulated as a linear programming model with an objective, cost function.

A classification of parts and machine tool components from a machine tool producer point of view is proposed. This classification is based on statistical analysis and technological and form similarity of parts. It results in a statistical determination of the market demands of machine tools.

Calculation methods for the stiffness of principal structural elements of a machine tool are developed.

A cost evaluation method is discussed. Finally, the derivation of a numerical figure of merit, representing the design and economic merits of a machine tool and its components, is described.

Also, methods of controlling and improving the structural components of machine tools are discussed, stressing the potential possibilities of the control of the damping characteristics rather than that of the "stiffness/mass" ratio.

A new concept of a dynamic evaluation criterion is proposed and its integration in the developed numerical complex design criterion and numerical figure of merit concepts is discussed.

Simplified numerical examples, based on the developed computation techniques for the determination of the deformation of structural elements under load, and on cost figures obtained from a large U.S. machine-tool manufacturer, are presented.

The concept of the numerical figure of merit demonstrates its value even in its over-simplified form.

THE IMPACT OF TECHNOLOGY ON THE MENTAL
CONTENT OF WORK IN INDUSTRIAL OPERATIONS

John Peter van Gigch, Ph.D.
Oregon State University, 1968

Industrial operations are becoming more mechanized and automated. The increasing complexity of work affects the proportion of physical to mental effort which workers spend in the performance of their jobs. Traditional work measurement techniques which are primarily designed to measure the individual's physical load have to be complemented with other techniques to evaluate his mental load. This study is aimed at developing a methodology which can be applied to the analysis and measurement of the mental contribution of workers and its associated effects.

Information Theory and the Theory of Communication are used to provide a conceptual framework in which information processes employed in the performance of industrial tasks are identified and evaluated.

Depending on their complexity, the various information processes called mental therbligs, are classified according to four levels of integrative behavior. Information processing rates are calculated on the basis of the amount of information transmitted in the performance of these therbligs. Average and, where possible, peak rates are obtained for typical jobs representing two or more technological levels in four different segments of the forest products industry: a. Lumber Sorting, b. Lumber Grading, c. Groundwork Pulp Production, and d. Sulfate (Kraft) Pulp Production. Thus, the mental content of jobs in each of these industry segments is evaluated.

The mental contribution of industrial workers is then analyzed in terms of some of the following:

1. The effects of the repetition cycle rate and the variability of the tasks' sequence.

2. The effects of increased system complexity on the ability of operators to cope with the high informational load, equipment malfunctions and emergency situations.

3. The effects of increased system entropy on the operator's speed of responses and on the requirements of system design.

4. The effects of the addition of a process control computer on the variability of the process parameters and on the operator's mental load.

5. The implications regarding training and compensation of workers slated for the jobs created by new technology.

a supplementary view...

THE WRITING OF ABSTRACTS By *Christian K. Arnold*

From IRE Transactions on Engineering Writing and Speech, *EWS-4, No. 3, December, 1961. Reprinted by permission of The Institute of Electrical and Electronics Engineers, Inc. and the author.*

Summary—The abstract is not only the most important section of your report or paper, but also the most difficult to write. Effective abstracts (1) must contain enough specific information to satisfy the needs of a research worker looking for sources of information and of an administrator looking for a progress or status definition; (2) must be a complete, self-sufficient unit; (3) must be made as short as possible without violating accuracy or eliminating essential information; (4) must be written in fluent, easily understood language; (5) must be consistent in tone and emphasis with the parent report or paper; and (6) should make the widest possible use of numerals and standard, generally recognized abbreviations.

The most important section of your technical report or paper is the abstract. Some people will read your report from cover to cover; others will skim many parts, reading carefully only those parts that interest them for one reason or another; some will read only the introduction, results, and conclusions; but everyone who picks it up will read the abstract. In fact, the percentage of those who read beyond the abstract is probably related directly to the skill with which the abstract is written. The first significant impression of your report is formed in the reader's mind by the abstract; and the sympathy with which it is read, if it is read at all, is often determined by this first impression. Further, the people your organization wants most to impress with your report are the very people who will probably read no more than the abstract and certainly no more than the abstract, introduction, conclusions, and recommendations. And the people you should want most to read your paper are the ones for whose free time you have the most competition.

Despite its importance, you are apt to throw your abstract together as fast as possible. Its construction is the last step of an arduous job that you would rather have avoided in the first place. It's a real relief to be rid of the thing, and almost anything will satisfy you. But a little time spent in learning the "rules" that govern the construction of good abstracts and in practicing how to apply them will pay material dividends to both you and your organization.

The abstract—or summary, foreword, or whatever you call the initial thumbnail sketch of your report or paper—has two purposes: (1) it provides the specialist in the field with enough information about the report to permit him to decide whether he could read it with profit, and (2) it provides the administrator or executive with enough knowledge about what has been done in the study or project and with what results to satisfy most of his administrative needs.

It might seem that the design specifications would depend upon the purpose for which the abstract is written. To satisfy the first purpose, for instance, the abstract needs only to give an accurate indication of the subject matter and scope of the study; but, to satisfy the second, the abstract must summarize the results and conclusions and give enough background information to make the results understandable. The abstract designed for the first purpose can tolerate any technical language or symbolic shortcuts understood at large by the subject-matter group; the abstract designed for the second purpose should contain no terms not generally understood in a semitechnical community. The abstract for the first purpose is called a *descriptive abstract*; that for the second, an *informative abstract*.

The following abstract, prepared by a professional technical abstracter in the Library of Congress, clearly gives the subject-matter specialist all the help he needs to decide whether he should read the article it describes:

Results are presented of a series of cold-room tests on a Dodge diesel engine to determine the effects on starting time of (1) fuel quantity delivered at cranking speed and (2) type of fuel-injection pump used. The tests were made at a temperature of $-10°F$ with engine and accessories chilled at $-10°F$ at least 8 hours before starting.

Regardless of however useful this abstract might be on a library card or in an index or an annotated bibliography, it does not give an executive enough information. Nor does it encourage everyone to read the article. In fact, this abstract is useless to everyone except the specialist looking for sources of information. The descriptive abstract, in other words, cannot satisfy the requirements of the informative abstract.

But is the reverse also true? Let's have a look at an informative abstract written for the same article:

A series of tests was made to determine the effect on diesel-engine starting characteristics at low temperatures of (1) the amount of fuel injected and (2) the type of injection pump used. All tests were conducted in a cold room maintained at $-10°F$ on a commercial Dodge engine. The engine and all accessories were "cold-soaked" in the test chamber for at least 8 hours before each test. Best starting was obtained with 116 cu mm of fuel, 85 per cent more than that required for maximum power. Very poor starting was obtained with the lean setting of 34.7 cu mm. Tests with two different pumps indicated that, for best starting characteristics, the pump must deliver fuel evenly to all cylinders even at low cranking speeds so that each cylinder contributes its maximum potential power.

This abstract is not perfect. With just a few more words, for instance, the abstracter could have clarified the data about the amount of fuel delivered; do the figures give flow rates (what is the unit of time?) or total amount of fuel injected (over how long a period?). He could easily have defined "best" starting. He could have been more specific about at least the more satisfactory type of pump: what is the type that delivers the fuel more evenly? Clarification of these points would not have increased the length of the abstract significantly.

The important point, however, is not the deficiencies of the illustration. In fact, it is almost impossible to find a perfect, or even near perfect, abstract, quite possibly because the abstract is the most difficult part of the report to write. This difficulty stems from the severe limitations imposed on its length, its importance to the over-all acceptance of the report or paper, and, with informative abstracts, the requirement for simplicity and general understandability.

The important point, rather, is that the informative abstract gives everything that is included in the descriptive one. The informative abstract, that is, satisfies not only its own purpose but also that of the descriptive abstract. Since values are obtained from the informative abstract that are not obtained from the descriptive, it is almost always worth while to take the extra time and effort necessary to produce a good informative abstract for your report or memo. Viewed from the standpoint of either the total time and effort expended on the writing job as a whole or the extra benefits that accrue to you and your organization, the additional effort is inconsequential.

It is impossible to lay down guidelines that will lead always to the construction of an effective abstract, simply because each reporting job, and consequently each abstract, is unique. However, general "rules" can be established that, if practiced conscientiously and applied intelligently, will eliminate most of the bugs from your abstracts.

1. *Your abstract must include enough specific information about the project or study to satisfy most of the administrative needs of a busy executive.* This means that the more important results, conclusions, and recommendations, together with enough additional information to make them understandable, must be included. This additional information will most certainly include an accurate statement of the problem and the limitations placed on it. It will probably include an interpretation of the results and the principal facts upon which the analysis was made, along with an indication of how they were obtained. Again, *specific* information must be given. One of the most common faults of abstracts in technical reports is that the information given is too general to be useful.

2. *Your abstract must be a self-contained unit, a complete report-in-miniature.* Sooner or later, most abstracts are separated from the parent report, and the abstract that cannot stand on its own feet independently must then either be rewritten or will fail to perform its job. And the rewriting, if it is done, will be done by someone not nearly as sympathetic with your study as you are. Even if it is not separated from the report, the abstract must be written as a complete, independent unit if it is to be of the most help possible to the executive. This rule automatically eliminates the common deadwood phrases like "this report contains..." or "this is a report on..." that clutter up many abstracts. It also eliminates all references to sections, figures, tables, or anything else contained in the report proper.

3. *Your abstract must be short.* Length in an abstract defeats every purpose for which it is written. However, no one can tell you just how short it must be. Some authorities have attempted to establish arbitrary lengths, usually in terms of a certain percentage of the report, the figure given normally falling between three and ten per cent. Such artificial guides are unrealistic. The abstract for a 30-page report must necessarily be longer, percentagewise, than the abstract for a 300-page report, since there is certain basic information that must be given regardless of the length of the report. In addition, the information given in some reports can be summarized much more briefly than can that given in other reports of the same over-all dimensions. Definite advantages, psychological as well as material, are obtained if the abstract is short enough to be printed entirely on one page so that the reader doesn't even have to turn a page to get the total picture that the abstract provides. Certainly, it should be no longer than the interest span of an only mildly interested and very busy executive. About the best practical advice that can be given in a vacuum is to make your abstract as short as possible without

cutting out essential information or doing violence to its accuracy. With practice, you might be surprised to learn how much information you can crowd into a few words. It helps, too, to learn to blue-pencil unessential information. It is perhaps important to document that "a meeting was held at the Bureau of Ordnance on Tuesday, October 3, 1961, at 2:30 P.M." somewhere, but such information is just excess baggage in your abstract: it helps neither the research worker looking for source material nor the administrator looking for a status or information summary. Someone is supposed to have once said, "I would have written a shorter letter if I had had more time." Take the time to make your abstracts shorter; the results are worth it. But be careful not to distort the facts in the condensing.

4. *Your abstract must be written in fluent, easy-to-read prose.* The odds are heavily against your reader's being an expert in the subject covered by your report or paper. In fact, the odds that he is an expert in your field are probably no greater than the odds that he has only a smattering of training in any technical or scientific discipline. And even if he were perfectly capable of following the most obscure, tortured technical jargon, he will appreciate your sparing him the necessity for doing it. T. O. Richards, head of the Laboratory Control Department, and R. A. Richardson, head of the Technical Data Department, both of the General Motors Corporation, have written that their experience shows the abstract cannot be made too elementary: "We never had [an abstract] . . . in which the explanations and terms were too simple." This requirement immediately eliminates the "telegraphic" writing often found in abstracts. Save footage by sound practices of economy and not by cutting out the articles and the transitional devices needed for smoothness and fluency. It also eliminates those obscure terms that you defend on the basis of "that's the way it's always said."

5. *Your abstract must be consistent in tone and emphases with the report proper, but it does not need to follow the arrangement, wording, or proportion of the original.* Data, information, and ideas introduced into the abstract must also appear in the report or paper. And they must appear with the same emphases. A conclusion or recommendation that is qualified in the report proper must not turn up without the qualification in the abstract. After all, someone might read both the abstract and the report. If this reader spots an inconsistency or is confused, you've lost a reader.

6. *Your abstract should make the widest possible use of abbreviations and numerals, but it must not contain any tables or illustrations.* Because of the space limitations imposed upon abstracts, the rules

governing the use of abbreviations and numerals are relaxed for it. In fact, all figures except those standing at the beginning of sentences should be written as numerals, and all abbreviations generally accepted by such standard sources as the American Standards Association and "Webster's Dictionary" should be used.

By now you must surely see why the abstract is the toughest part of your report to write. A good abstract is well worth the time and effort necessary to write it and is one of the most important parts of your report. And abstract writing probably contributes more to the acquisition of sound expository skills than does any other prose discipline.

Part III
THE TECHNICIAN WRITES FUNCTIONAL ENGLISH

Chapter 15
FUNCTIONAL ENGLISH

The technician will be familiar with the word "functional" in a variety of contexts. We speak of functional design or of functional architecture, for example. By this we refer to the relationship between the way a thing is made and the special purpose or use for which it is intended. A table is functional if it can be used for eating. If the table top can be raised to another height and tilted, it might also function as a drawing board. A table constructed to fulfill two specific purposes—eating and drawing—is functional in that its design grows out of the characteristic uses to which it will be put.

Functionalism is the adaptation of form or structure to activity or use. Functional English is adapted by form and structure to a special use. In technical writing the function of English is to convey facts clearly, logically, simply, and objectively.

WORDS IN FUNCTIONAL ENGLISH

The language of science, functional English, is the language of the technician writing a report. The primary intention of the technical writer is not to argue, not to describe, not to narrate, but to explain factual information.

The writer is first of all concerned with the correct choice of individual words. A word has an actual meaning which we say is its *denotation*. But beyond its actual meaning a word may suggest or imply other meanings through associations carried in a reader's mind. An implied meaning is the *connotation* of the word. In writing functional English the technician tries to use words which do not arouse feelings that may interfere with the explanation at hand. A writer knows, for example, of many occasions when he might use such words as *labor agitator* or *riot*. But where his purpose as a writer is to explain facts objectively, he avoids words that are charged with emotional associations and open to interpretation in favor of words with relatively precise impersonal meanings or denotations.

Perhaps no student more than the technical student is aware of the special vocabulary peculiar to science and technology. Each field of study in school has its own terms, its own words, but technical words may present difficulties not only for non-technicians but for technicians as well. At the same time, it is of great importance that the technical man use words accurately and precisely.

In an earlier section of this book dealing with the resources of the library the student found listed some of the many general and special reference works at the disposal of the technical man. None can be more helpful as a basic tool than the dictionary or handbook of a specific scientific field or technical discipline.

It has often been said that a technical writer must be a writer first and a technician second. Ideally the technical writer would be in possession of general writing principles which he could apply in writing of his specific technology. There are several principles of writing which apply generally but which the technician will find useful as he brings them to bear upon his particular writing problems. The principles deal with words, sentences, and paragraphs.

One practice concerns general-abstract and concrete-specific words. The word *vehicle*, for example, refers generally to that entire class of objects which are used as the instrument of conveyance of things or people. To arrive at the meaning of the word *vehicle* the reader must abstract or leave out specific characteristics of a covered wagon or jet airliner and concentrate on characteristics common to all vehicles. As a word referring to a general class of objects, *vehicle* is an abstract term relatively devoid of specific characteristics. A word is said to be general when, like *vehicle*, it refers to a general set of things, persons, places, or acts.

Because of its general-abstract character, the word *vehicle*—which denotes a whole class of things—will imply multiple different characteristics and qualities to different readers. There will be a whole range of different understandings and, therefore, a great possibility of vague or incorrect communication where the word is used in a report.

Vehicle is general-abstract—*automobile* is also, but relatively less so. A word, it may easily be seen, is general-abstract in relative degree—that is, *automobile* is more concrete-specific than *vehicle*. While *automobile* is less general than *vehicle*, *Cadillac* is more concrete than either of the other two terms, and *Black Cadillac Model A 3 with white walled tires and power steering* is even more so. There are clearly relative degrees of general-abstract. But the more concrete-specific a word is, the less it obliges the reader of a technical communication to abstract or leave out general characteristics common only to a whole class of things such as vehicles. If the reader is left free to concentrate fully on the concrete—specific qualities of Cadillac Model A 3, he is more likely to arrive at the meaning intended by the writer. In technical writing, the chances for clearer communication are increased as the writer chooses the relatively more concrete-specific words available to him.

General-abstract language is highly conceptualized; it tends to refer to abstract concepts and ideas and is therefore more suited to pure or theoretical science than it is to technology. The basic concern of the new technical writer is for facts and their fundamental relationships. Concrete-specific words refer more definitely to the individual things for which they stand and are thus more likely to serve the needs of technical readers and writers for precision and definiteness.

STYLE IN FUNCTIONAL ENGLISH

A discussion of the need for concrete-specific words in functional English naturally leads to the larger question of style in technical writing. We speak of style of dressing, style of dancing, style of architecture, and style of writing. The word *style* refers to the same idea in all cases—the sum total of the distinctive characteristics and attributes which make an idea or object or piece of writing one thing and not another. Our concern is for the distinctive style of functional English used in technical writing.

The aim of functional English, as we said earlier, is to be clear and concise. But how does a beginning technical writer achieve clarity and conciseness in his style? As to being clear, we have already suggested the use of concrete-specific words in preference to general-abstract terms.

The great literature of the world—the works of Chaucer, Shakespeare, and Milton in English, for example—is imaginative literature. It is literature that might be said to aim at awakening self-knowledge in the reader. The great imaginative writer functions as a kind of superb magician who confers on words awesome powers that somehow touch the spirit of men. As great artists and writers of genius, creators of imaginative literature use all the devices of language possible to awaken the spirits of readers. There will be words rich with connotations and overtones of meaning; there will be imagery, metaphor, and symbolism; there will be understatement, overstatement, play on words, and startling descriptive passages. The imaginative writer works artfully by suggestion, subtlety, indirection, implication, inference, and surprise to engage the whole imaginative and spiritual character of the reader.

As imaginative literature has a glorious history of development in the western world, so does scientific literature. In order to appreciate fully the appropriate style for technical writing, the student should ideally have knowledge of the writings of the great scientific thinkers and stylists who laid the groundwork for science in the modern world. If technicians are to communicate intelligently and work directly with scientists, engineers, and other professionals in every field of science and technology, they should have at least a broad familiarity with the great scientific-theoretical literature of our civilization; and they should have more than a superficial knowledge of the method of scientific inquiry. For example, the foreground of modern science and technology may be seen in the writing of the following men. How many of these names do you know at first hand from their writings—Aristotle, Hippocrates, Galen, Euclid, Archimedes, Ptolemy, Copernicus, Kepler, Galileo, Harvey, Bacon, Descartes, Newton, Faraday, and Darwin?

a supplementary view...

WHAT CAN THE TECHNICAL WRITER OF THE PAST TEACH THE TECHNICAL WRITER OF TODAY? By *Walter James Miller*

From IRE Transactions on Engineering Writing and Speech, *EWS-4, No. 3, December, 1961. Reprinted by permission of the Institute of Electrical and Electronics Engineers, Inc. and the author.*

In discussing the question of what we can learn from the "technical writer" of the past, I do not propose to take you on one of those outline-of-history tours which quickly become foggy blurs of names and dates. Anything like a detailed history of technical writing would involve a discussion of so many literary works that most of them would—at least in the time allotted to us here—get mere mention in passing. Worse yet, a thorough historical study would involve consideration of many technical writings that have exerted tremendous historical and social influence but have enjoyed very little intrinsic literary merit.

For example, we would be considering Dr. Frederick W. Taylor's "Scientific Management" which, it is true, revolutionized American managerial philosophy and techniques; Benjamin Garver Lamme's "Electrical Engineering Papers" which, it is true, organized electrical engineering into teachable structures of ideas; and the 1798 edition of "Encyclopaedia Britannica" which, it is a fact, was the working Bible of early American applied science. But all of these works hold no literary interest for the technical writer today, except to drive home, *ad nauseam*, the point that he can accomplish miracles simply by working from a good outline. The history of scientific and technical writing is replete—like the history of the literatures of esthetics and music—with numerous works valuable for their content, important to their day, but very very feeble in literary quality.

No, for our needs it would be wiser to consider only the highlights—or even only some of the highlights—of the history of scientific and technical writing. Given, then, all the historical studies and translations by Drachman, Hoover, Herschel, Shapley, Finch, the Newcomen Society, Miller and Saidla, to name a few, we ask simply, "What comes

227

out of these researches that is of value to the man more interested in writing than in history?"

To work toward an answer, and to spark our discussion, I shall limit myself to twelve important writers, dating from ancient times to now, each important not only for what he said, but for how he said it. And by limiting myself in this manner, I shall be able to illustrate my discussion of literary techniques with meaty quotations, so that we may appreciate some of the literature as well as some of the history. And, for my conclusion, I hope to garner some lessons that we can learn from these classic works. In other words, I hope to sell these authors to you, to excite you to read them more thoroughly at your convenience.

In selecting writers to make my points, I shall go back over the main line of descent of the modern American technical writer. Like any American writer, our technical writer is descended from the British, tracing his literary ancestry back to the Continent and thence to the ancient world. Starting at that far end, then, I shall talk about Vitruvius and Frontinus, as representative of ancient technical literature; Biringuccio and Agricola, as representative of Renaissance Italian and German literature; Smeaton, McAdam, and Rankine, as representing early British technical writing; Eads and Wellington, early American; and Hoover, Ammann, and Raymond, contemporary American.

One might protest the absence of Archimedes, Francis Bacon, Hugh Miller, Charles Darwin, or Albert Einstein. I would reply that these writers are already familiar to us, and that I want to make my points while introducing you to relatively unknown, still unappreciated, but also important figures in history of technical writing. Further, I want to make my points in terms of writers in the applied sciences, a field in which most technical writing is being done today and in which the problems of raising the literary level are most urgent.

1. Vitruvius

The first important technical writer that history commends to our attention is Marcus Vitruvius Pollio, architect in the service of Augustus Caesar. In about 27 B.C. Vitruvius faced a literary problem slightly greater than that faced today by a technical writer preparing a manual on the maintenance of an IBM machine. Vitruvius was assigned to write a treatise for the master builders of Rome, to provide them with a thorough theoretical and practical guide to city planning, civil and military structures, interior decoration, hydraulics, instruments, tools, machines, engines, and all known related subjects. His "Ten Books on Architecture" served as the authority for generations of Romans and their descendants, including Leonardo da Vinci, and was still consulted eighteen centuries after it was written by such prominent engineers as the Englishman, Thomas Telford.

Of course, on such topics as the nature of materials, Vitruvius is now out of date, but it is astonishing how much he can still teach us about writing. He employed successfully all the literary devices known to the conscientious author. Let us read, for instance, a few sentences from his eloquent description of the profession of architecture:

> The architect should be equipped with knowledge of many branches of study and varied kinds of learning.... This knowledge is the child of practice and theory. Practice is the continuous and regular exercise of employment where manual work is done according to the design of a drawing. Theory... is the ability to demonstrate and explain the productions of dexterity on the principles of proportion.

Several hundred words later he sums up thus:

> since this study is so vast in extent, embellished and enriched as it is with many different kinds of learning, I think that men have no right to profess themselves architects hastily, without having climbed from boyhood the steps of these studies and thus, nursed by the knowledge of many arts and sciences, having reached the heights of the holy ground of architecture.

Note, first of all, the excellent use of metaphor, ranging from the simple "knowledge is the child of practice and theory" to the reverential "heights of the holy ground of architecture." Note, too, that Vitruvius buckles down in as early as his third and fourth sentences to definitions of such easily-taken-for-granted terms as "theory" and "practice."

Flipping a few pages into the technical treatise proper, I read:

> The town being fortified, the next step is... the laying out of streets and alleys with regard to climatic conditions. They will be properly laid out if foresight is employed to exclude the winds.... Cold winds are disagreeable, hot winds enervating, moist winds unhealthful. We must, therefore, avoid mistakes and beware of the experience of many communities. For example, Mytilene in the island of Lesbos is a town built with magnificence and good taste, but its position shows a lack of foresight. In that town when the wind is south, the people fall ill; when it is northwest, it sets them coughing; with a north wind they do indeed recover but cannot stand about in the streets owing to the severe cold.

Could there be, in all the literature of technics, a better example of a good example?

Let me make my final points about Vitruvius from his well-moulded description of pozzolana cement:

> There is a kind of powder which from natural causes produces astonishing results. It is found in the neighborhood of Baiae and around Mount Vesuvius. This substance, when mixed with lime and rubble, not only lends strength to buildings of other kinds, but even when piers of it are constructed in the sea, they set hard under water. The reason for this seems to be that the soil on the slopes of the mountains in these neighborhoods is hot and full of hot springs. This would not

be so unless the mountains had beneath them huge fires of burning sulfur or alum or asphalt. So the fire and the heat of the flames, coming up from far within the fissures, make the soil there light, and the tufa found there is spongy and free from moisture. Hence, when the three substances, all formed on a similar principle by the force of fire, are mixed together, the water suddenly taken in makes them cohere, and the moisture quickly hardens them so that they set into a mass which neither the waves nor the force of the water can dissolve.

In all his descriptions Vitruvius manages to communicate his wonderment, his excitement over the materials with which he works. But in his excitement he never forgets the needs of his reader. He says, later on:

There will still be the question why Tuscany, although it abounds in hot springs, does not furnish a powder out of which, on the same principle, a wall can be made which will set fast under water. I have therefore thought best to explain how this seems to be, before the question can be raised.

In his careful anticipation of questions likely to occur to his reader, in his use of functional metaphors, timely definitions, superb examples, and in his descriptions which catch the technical man's love of his art, Vitruvius is truly an inspiration even to the technical writer of 1961.

2. Frontinus

Another Roman who can, across centuries, give us perspective on our writing problems is Sextus Julius Frontinus. When he took over as water commissioner of Rome in 97 A.D., he found himself to be heading a department wallowing in sinecure and graft, while 40 per cent of the water supply was lost or stolen, new conduits were in disrepair, some aqueducts were bringing contaminated water into the city, and the most crowded sections were suffering droughts. As further evidence of low morale and inefficiency in his department, he discovered that there were no adequate plans of the works, no comprehensive treatise on the operation and administration of the water system.

Then in his sixties, Frontinus rolled up his sleeves. In a few years Rome's water supply was doubled, its water and even its air were purer. Water engineers worked from master plans and from an excellent information report prepared for them by Frontinus. He had found himself in a situation in which he sensed the need for—and apparently invented—this (now indispensable) technical genre, the *information report*. In his "Aqueducts of Rome" he tackled some of the timeless problems of this type. For example, he knew what many technical writers still do not know, that readers of reports will differ tremendously in the degree and extent of their technical background and interest, that some readers want only a general picture, while others want to luxuriate in details. And so, the considerate Frontinus set up alternate routes through his report, advising the less technical-minded readers when and how they could make a quick detour around a jungle of thorny data.

In style Frontinus is equally imaginative. His metaphors are bolder, more concrete than those of Vitruvius. He sees water rushing in from the aqueducts and slowing down in the catch-basins as "taking fresh breath... after the run." He sees his assistants as "but the hands and tools of the directing head." Uninhibited by modern "objectivity," he can, first of all, spice his work with sly sarcasm, usually at the expense of administrators less conscientious than himself. Note here how he plays it absolutely dead-pan:

> I deem it of the greatest importance to familiarize myself with the business I have undertaken. For... there is nothing so disgraceful for a decent man as to conduct the office delegated to him, according to the instructions of assistants.

He can, secondly, indulge in pride in his work, as when he looks at his 240 miles of aqueducts and observes, smugly:

> With such an array of indispensable structures carrying so many waters, compare if you will the idle pyramids or the useless though famous works of the Greeks!

3. Biringuccio

For fifteen centuries after these Roman works were composed, we find very little technical writing of interest to us; science and technics were, during the Middle Ages, in relative disrepute, and hence, comatose. As a matter of fact, when technical books were first printed in the early Renaissance, publishers had no contemporary manuscripts of any significance on hand, and Vitruvius and Frontinus were the best sellers, both in Latin and in translation.

The earliest significant works in modern technical literature, from our point of view at least, were in the fields of mining and metallurgy. During the sixteenth century, manuals on digging, smelting, and alloying were in great practical demand. Competition among authors helped produce, finally, two superior books still of interest to us today.

In 1540 Vannoccio Biringuccio published "De La Pirotechnica," a work instructive to us for the way in which enthusiasm and love of subject can permeate the most technical information. Some moderns would say that Biringuccio's enthusiasm has gone too far, because to his eager chapters on the fires of metallurgy he added an eager chapter on the fires of love, "the fire," he observes, in one of my favorite sentences in all western literature, "that consumes without leaving ashes."

Obviously Biringuccio was writing before narrow specialization set in, before the technical man saw any separation between work and life. But it is precisely in this that Biringuccio, it seems to me, has an important message for the modern technical writer. Biringuccio could write with passion about metallurgy because he did not confine his passion to metallurgy.

4. Agricola

The other writer of this period still qualified to teach us something about the writing craft is Georgius Agricola, a German medical doctor who wrote voluminously on the plague, weights and measures, geology and other subjects. His *magnum opus* was "De Re Metallica," published in 1556, which reigned as the standard text on the mining and working of metals for nearly two centuries. For us this treatise is itself an inexhaustible mine of good examples of technical description and scientific argumentation.

In most of his descriptions of apparatus Agricola uses the simple but dramatic techniques of starting with a small central part—a stick or a brick—and adding to it piece by piece, until gradually the static structure, *e.g.*, a hauling machine or a furnace, is a scene of activity, giving off its working noises, smells, and results. Agricola was well aware of the value of sense appeal, especially in describing a process. Notice how important are the cues for hand, eye, and ear in this passage:

first of all they hew out the rock of the hangingwall or of the footwall if it be less hard; then they place timbers set in hitches in the hanging or footwall, a little above the vein, and from the front and upper part, where the vein is seen to be seamed with small cracks, they drive into one of the little cracks one of the iron tools I have mentioned; in each fracture they place four thin iron blocks, and in order to hold them more firmly, if necessary, they place as many thin iron plates back to back; next they place thinner iron plates between each two iron blocks, and strike and drive them by turns with hammers, whereby the vein rings with a shrill sound; and the moment when it begins to be detached from the... wall rock, a tearing sound is heard. As soon as this grows distinct the miners hastily flee away; then a great crash is heard as the vein is broken and torn and falls down.

To his exploitation of the five senses, add another reason for admiring his descriptions. Although Agricola's book was illustrated with hundreds of woodcuts, he rarely referred to them in his text and never depended on them to make his point. He accepted the full literary challenge of making his writing a self-explanatory medium.

In his writing Agricola could build an idea as easily as he could build a furnace. His defense of the metallic arts, with which he opens his book, is one of the great intellectual efforts of science in its struggle against ignorance and superstition. It ranks with the best defenses in modern literature of evolution and vivisection. Agricola sympathetically entertains, even forcefully restates, each argument of his opposition (*e.g.*, "it is contrary to Nature to dig up what God has concealed") before he answers it with patience and often with cleverness ("then why fish?").

After pulverizing his enemies' arguments with logical analysis, he caps his defense with a description of man without metals, inept and inhibited in farming, sewing, building:

> If we remove metals from the service of man, all methods of protecting and sustaining health and more carefully preserving the course of life are done away with. If there were no metals, men would pass a horrible and wretched existence in the midst of wild beasts; they would return to the acorns and fruits and berries of the forest. They would feed upon the herbs and roots which they plucked up with their nails. They would dig out caves in which to lie down at night, and by day would rove in the woods and plains at random like beasts, and inasmuch as this condition is utterly unworthy of humanity, with its glorious and splendid natural endowment, will anyone be so foolish or obstinate as not to allow that metals are necessary for food and clothing and that they tend to preserve life?

To his successful techniques of exposition, add still another techique instructive to us—his technique of research. Agricola was one of the first technical writers to stuff his pride and university degrees into his pocket and to interview the working technician for what he could contribute.

5. Smeaton

Nothing comparable to these works appears in the English literature of applied science for many generations. For the Renaissance reached England, on Europe's edge, last. English was the last major European language to produce a translation of Vitruvius.

In 1754, when the distinguished scientist John Smeaton decided to specialize in engineering, he could not find enough systematic writing in English to educate himself. He had to learn French and study the Continental masters. But Smeaton remedied this situation. After a long life of work on canals, mills, and engines, he undertook to rewrite his reports and notes into organized commentaries. His "Narrative of the Building of the Eddystone Lighthouse," planned as the first of a series of narratives, appeared in 1791. After his death, his colleagues carried out his project as best they could, publishing his "Reports" and "Papers" in four fat volumes. Smeaton's five books covered such a great range of problems and projects that they served as textbooks for a generation. Smeaton had founded the English literature of engineering.

His "Reports" are famous for the way he discusses profound technical subjects in the simplest language. His "Narrative" develops an epic adventure in warm, poetic language. Smeaton elicits sympathy, creates suspense, as he describes his struggle to root his lighthouse on the notorious reef, and describes his feelings when his ship is lost in a storm, his workers balk, his writing humbles him with its difficulties. But metaphor and analogy come readily to him. Stone set in the frame of a building acts "like the ballast of a ship." The narrowness of a building's base "rounds it like the rockers of a cradle." And, if he does not plan carefully, building the Eddystone Lighthouse will be like "the rolling of the stone of Sisyphus."

His planning and his efforts to explain his planning to his reader develop partly in terms of a poetic, but nevertheless scientific, analogy. Here is a portion of his account of how he conceived the proper shape for a stone lighthouse:

On this occasion, the natural figure of the waist or bole of a large spreading oak presented itself to my imagination. Let us for a moment consider this tree: suppose at twelve or fifteen feet above its base, it branches out in every direction, and forms a large bushy top. . . . This top, when full of leaves, is subject to a very great impulse from the agitation of violent winds; yet partly by its elasticity, and partly by the natural strength arising from its figure, it resists them all, even for ages, till the gradual decay of the material diminishes the coherence of the parts, and they suffer piecemeal by the violence; but it is very rare that we hear of such a tree being torn up by the roots.

Let us now consider its particular figure. Connected with its roots, which lie below ground, it rises from the surface thereof with a large swelling base, which at the height of one diameter is generally reduced by an elegant curve, concave to the eye, to a diameter less by at least one-third, and sometimes to half of its original base. From thence its taper diminishing more slow, its sides by degrees come into a perpendicular, and for some height form a cylinder. After that a preparation of more circumference becomes necessary, for the strong insertion and establishment of the principal boughs, which produces a swelling of its diameter.

Now we can hardly doubt but every section of the tree is nearly of an equal strength in proportion to what it has to resist; and were we to lop off its principal boughs, and expose it in that state to a rapid current of water, we should find it as much capable of resisting the action of the heavier fluid, when divested of the greatest part of its clothing, as it was that of the lighter when all its spreading ornaments were exposed to the fury of the wind: and hence we may derive an idea of what the proper shape of a column of the greatest stability ought to be, to resist the action of external violence. . . .

Thus, the now classic torso of the stone lighthouse evolved in an extended comparison.

6. McAdam

The next British writer recommended as a stimulating influence for the modern is Thomas Loudon McAdam. The United States, incidentally, might have claimed him as one of its major writers and practitioners in applied science if it had not been for the intolerance of those New Yorkers who, after the American Revolution, just could not make the Tory McAdam feel at home on this side of the Atlantic. And so he macadamized the roads of Britain and Europe first, America welcoming his system long after it had unwelcomed its inventor.

McAdam's best work is his classic "Remarks on the Present System of

Road-Making," which contains the crux of the McAdam theory. The function of a road is not, as his predecessors had believed, to carry the traffic. Rather, the purpose of a road is to keep the ground beneath it dry so that the earth itself can bear the burden. Note how again metaphor helps make the writer. A McAdam road, he says, is "a roof." His competitors' roads are "reservoirs for water."

McAdam's writing is just like the original McAdam road. His words are sharp, angular stones, pressed together into a hard smooth surface, easy to travel on, and leading swiftly to the goal.

7. Rankine

Probably the greatest applied-science writer in the English language was William J. M. Rankine. In 150 papers and four treatises he transformed engineering from an art to a science. Organizing, and in some areas creating, its theory, he treated it on the highest philosophical, mathematical, and scholarly level it had so far reached. For this reason, he was ridiculed by the giants of the "rule of thumb" era. Nevertheless, Rankine's books enjoyed a phenomenal career. The 13th edition of his "Manual of the Steam Engine and Other Prime Movers" was published in 1891, the 23rd edition of "Manual of Civil Engineering" in 1908, and the 21st edition of "Manual of Applied Mechanics" in 1921, 63 years after initial publication.

Reading Rankine today is still one of the supreme experiences in logical development of ideas. With unerring syntax, he handles ponderous masses of material in elegant fashion.

He is especially instructive in his masterful definitions. Whether defining a basic part of an engine, or dismantling a fallacious concept and building a correct one in its place, Rankine achieves a clean, simple, inevitable pattern of language. This is typical of hundreds of his now classic definitions of components of machinery. Note how he makes us visualize the components before he gives them names—the inductive and more suspenseful approach:

The part of a heat engine in which the fluid performs work consists essentially of an enclosed space whose volume is capable of being alternately enlarged and contracted by the motion of one of its boundaries. The enclosed space is of a cylindrical form, in all engines that are extensively used in practice; and it is called the *cylinder*, even in those exceptional engines in which it has some other figure. Its movable boundary is called the *piston*, and is usually a cylindrical disc fitting the cylinder in which it moves to and fro in a straight line. In some. . . engines the piston has other forms, but its action always is to increase and diminish alternately the volume of a certain enclosed space.

His greatest definition, very important in the history of ideas, is given in this famous passage:

> In the history of mechanical art two modes of progress may be distinguished—the *empirical* and the *scientific*. Not the *practical* and *theoretic*, for that distinction is fallacious: all real progress in mechanical art, whether theoretical or not, must be practical. The true distinction is this: that the empirical mode of progress is purely and simply practical; the scientific mode of progress is at once practical and theoretic.

In this sharp rearrangement of hitherto fuzzily focused concepts, he follows through with a fuller discussion of each mode, and finally giving excellent historical illustration by contrasting Smeaton's empirical work on the old-type steam engine with Watt's great scientific advance. It adds up to one of the most successfully sustained and extended definitions in the English literature.

The modern technical writer, discouraged by the endless chore of definition, will find Rankine's numerous variations in approach an endless source of inspiration.

8. Eads

Through such works as the "Encyclopaedia Britannica" and McAdam's "Remarks," British technical writers exerted tremendous influence on early American practice. But by mid-nineteenth century the relation between the old country and the new was one of reciprocal influence. For example, American contributions to the theory of structures made by Squire Whipple and Herman Haupt found their rightful place in Rankine's syntheses of applied science. Soon American literature, with writers like James B. Eads and Arthur M. Wellington, was impressing a world-wide audience.

In 1868 Captain Eads turned a great defeat into a great triumph, mainly through his literary ability. His St. Louis River Bridge project had suffered every setback: floods, timid financing, opponents willing to exploit popular fear. When the bridge seemed just about doomed, Eads composed a magnificent 58-page report explaining, in the first part, "the plan of the structure, the principles involved in its construction, and reasons for its preference," avoiding "all technicalities not understood by every one." But in his second part he gave "all the scientific data, principles and formulae" in a way calculated to satisfy "the most critical engineer." We have here an early example of an excellent literary device for satisfying the needs of a mixed audience, now known as "the double report form." It is a more satisfactory solution than Frontinus' detour.

Debating in the best of Agricolan tradition, Eads neatly demonstrated why an arch was cheaper than a truss, and then dismantled his opposition's claim that "there was no engineering precedent for a span of 500 feet." First he showed that there had been spans of such length—his opponents had counted on popular ignorance of this—but, not willing to win on the basis of facts alone, Eads now bore down to the bedrock of principle:

> Suppose there had not been any engineering precedent? Must we admit that because a thing never has been done, it never can be, when our knowledge and judgment assure us that it is entirely practicable? This shallow reasoning would have defeated the laying of the Atlantic Cable; the spanning of the Menai Straits; and left the terrors of the Eddystone without their warning light. The Rhine and the sea would still be alternately claiming dominion over one-half the territory of a powerful kingdom, if this miserable argument had been allowed to prevail.

Financiers and engineers everywhere read the Eads report. Scientific publications praised it. Public confidence was rewon.

Even today, reading Eads is a boost to the morale of the writer who despairs over the task of translating scientific information into public action.

9. Wellington

On this side of the Atlantic the first major work in technical writing was "The Economic Theory of the Location of Railways," published in 1887, by Arthur M. Wellington. Principles set forth in this 930-page treatise have become part of the modern engineer's thinking; some of them are part of his language. For Wellington was a master of the memorable sentence. "It would be well," he declared early in his book,

> if engineering were less generally thought of, and even defined, as the art of constructing. In a certain important sense it is rather the art of not constructing; or, to define it rudely but not inaptly, it is the art of doing that well with one dollar, which any bungler can do with two after a fashion.

You will recognize that this statement, in streamlined form, has become a proverb of the profession. And like many proverbs, it is considered by many unreading people who repeat it frequently to be a choice example of native folk wisdom rather than the work of a conscious literary artist. And a conscious literary artist Wellington certainly was; such sentences are not occasional feats for him. Such dithyrambic sentences roar through the book. A few of them drop exhausted at the bottom of the next page, but most of them are self-recharging in their vitality and excitement. Note how he continues from the passage just read:

> There are, indeed, certain great triumphs of engineering genius—the locomotive, the truss bridge, the steel rail—which so rude a definition does not cover, for the bungler cannot attempt them at all; but such are rather invention than engineering proper. There is also in some branches of engineering, as in bridge-building, a certain other side to it, not covered by such a definition, which consists in doing that safely, at some cost or other, which the bungler is likely to try to do and fail.

He therefore in such branches who is simply able to design a structure which will not fall down may doubtless... be called an engineer....

But to such engineering as is needed for laying out railways, at least, the definition given is literally applicable, for the economic problem is all there is to it. The ill-designed bridge breaks down; the ill-designed dam gives way; the ill-designed boiler explodes; the badly-built tunnel caves in, and the bungler's bungling is betrayed. But a little practice and a little study of... geometry will enable anyone of ordinary intelligence without any engineering knowledge whatever in the larger sense to lay out a railway from almost anywhere to anywhere, which will carry the locomotive with perfect safety, and perhaps show no obtrusive defects under what is too often the only test—inspection after construction from the rear-end of a palace car. Thus, for such work, the healthful checks which reveal the bungler's errors to the world and to himself do not exist. Nature, unhappily, has provided no way for the locomotive to refuse to pass over an ill-designed railway as it refuses to pass over an ill-designed bridge.

Working on a grander scale than Frontinus and Eads, Wellington also tackled the problem of accommodating different levels of technical interest in the reader. He had his book set in three sizes of type. The reader who wants only a general treatment need read only the large-type sections. To sample some technical depth, a reader delves into the medium-size print. If he must be comprehensive, he reads even the small face. Each reader makes his own book. Most technical writers today, not concerned about where to shunt a flat-car, will be content with the preface and the introduction which present splendidly in large type the most vigorous voice in the American literature. But if any writer doubts that vigor in writing can reach down into the deepest technical material, let him read Wellington's small type.

10. Ammann

We come now to the contemporaries. I must confess that, although the weekly mail from the public relations office of D. B. Steinman used to exhort us to read Steinman's writings as a sort of full-time avocation, I have always recommended that we save some time also to enjoy the writings of Othmar Ammann, Herbert Hoover, and Arthur B. Raymond. This is not to disparage the work of Steinman—at least not that work on the level of his masterful paper on the "Design of the Mackinac Bridge for Aerodynamic Stability"—but just to resist his press agent's propaganda that Steinman was the only writer alive, in verse or prose.

Ammann, several of whose bridges have helped make New York a beautiful city, at least from the air and the water, is also a good designer and builder of beautiful reports. Now in his eighties, and still an active senior partner of Ammann and Whitney, he has written more than 100 major reports and papers. His "Tentative Report on the

Hudson River Bridge," submitted in 1926, exemplifies the kind of report that technical writers are always urged to write: the kind that comes to the point quickly. Ammann needs only three sentences to state the occasion, aim, scope, and method of attack in his investigation. His fourth sentence introduces his conclusions, which are presented in an order that outlines the body of his report. The "complete and detailed account" that follows ramifies at length each of the conclusions in turn.

Administrators need read only the opening sections to find all they need. Engineers can read the body for the full professional experience. Like the journalist's "inverted pyramid," the engineer's "double report" is a well-designed answer to the ancient problem of different levels of interest.

Ammann casts his ideas in long-span sentences that carry heavy loads with spring and verve. Like most of the major technical writers, he abhors the grey language, or the so-called "objectivity," of the petty technical mind, and honestly translates the quantitative into the qualitative and the "subjective." The data reveal an "intolerable traffic situation." The architect must do justice "to a charming landscape." Explaining why he advocates the suspension-type bridge in the setting of the Hudson River and its Palisades, he says:

A cantilever bridge, the nearest other possibility, would, with its dense and massive network of steel members, form a monstrous structure and mar forever the beauty of the natural scenery.

Perhaps, after reading Ammann, it will not be surprising to learn that, although in this report he asked for $100,000 to carry on his planning for the bridge, the New York and New Jersey legislatures pledged him $10,000,000.

11. Hoover

Hoover is not a graceful writer. Legendary by now are his struggles to pass English when he was majoring in geology at Stanford University. But he is an excellent example of what deep conviction can accomplish when geared to hard work. To understand what I mean, compare Frederick Taylor to Hoover. Both worked hard at mastering the craft of writing, but Taylor came up only with clarity, while Hoover comes up with emotional strength in everything he writes.

For example, consider Hoover's "Report on the Mississippi Flood," written when he was Secretary of Commerce in 1927. Called upon to explain this national catastrophe and to reassure the public it would not recur, Hoover produced an important piece of well-organized technical journalism. He used comparisons and figures of speech sparingly, but with effect. He saw the river as "caged," restless in its man-made "trough"; freed, it runs at a rate "ten times that of Niagara" over an area "nearly as large as Indiana." But we cannot return this area to "mosquito-swamps"; we must "learn to live with this river." These are the big, simple expressions of a man engaged in an elemental struggle with a force he respects, a man who is all thumbs at subtle embroidery, but strong in hammering out the larger structures of his ideas.

Hoover's description of his profession of engineering—which many people, friends and enemies, feel he should never have left—ranks with the noble descriptions of their sciences by Hippocrates, Vitruvius, and Agricola. Notice how genuine feeling comes through, despite the occasional badly jointed syntax or weak diction:

> It is a great profession. There is the fascination of watching a figment of the imagination emerge through the aid of science to a plan on paper. Then it moves to realization in stone or metal or energy. Then it brings jobs and homes to men. Then it elevates the standards of living and adds to the comforts of life. That is the engineer's high privilege.
>
> The great liability of the engineer compared to men of other professions is that his works are out in the open where all can see them. His acts, step by step, are in hard substance. He cannot bury his mistakes in the grave like the doctors. He cannot argue them into thin air or blame the judge like the lawyers. He cannot, like the architects, cover his failures with trees and vines. He cannot, like the politicians, screen his shortcomings by blaming his opponents and hope the people will forget. The engineer simply cannot deny he did it. If his works do not work, he is damned. . . .
>
> On the other hand, unlike the doctor his is not a life among the weak. Unlike the soldier, destruction is not his purpose. Unlike the lawyer, quarrels are not his daily bread. To the engineer falls the job of clothing the bare bones of science with life, comfort, and hope. No doubt as years go by people forget which engineer did it, even if they ever knew. Or some politician puts his name on it. Or they credit it to some promoter who used other people's money. . . . But the engineer himself looks back at the unending stream of goodness which flows from his successes with satisfactions that few professions may know. And the verdict of his fellow professionals is all the accolade he wants.
>
> With the industrial revolution and the advancement of engineers to the administration of industry as well as its technical direction, the governmental, economic and social impacts upon the engineers have steadily increased. Once lawyers were the only professional men whose contacts with the problems of government led them on to positions of public responsibility. From the point of view of accuracy and intellectual honesty the more men of engineering background who become public officials, the better for representative government.

Note that study of the classics of technical writing has obviously played a part in Hoover's development as an author. Present, among other influences, are some of the stage tricks of Frontinus, Agricola, and Wellington. Incidentally, Hoover is the editor and one of the translators of the English edition of Agricola's "De Re Metallica" and he is the author of "Principles of Mining,' a standard text in the first quarter of our century.

12. Raymond

The last writer I shall nominate for your inspirational reading is Arthur E. Raymond, Vice-President of Douglas Aircraft Company, and author of half a dozen major papers and lectures. He reached his due recognition when in 1951 the Royal Aeronautical Society invited him to give the Wilbur Wright Memorial Lecture. To 400 aviation experts gathered in London in the impressive setting of the Third Anglo-American Aeronautical Conference, Raymond offered a working philosophy of technical administration in a paper entitled "The Well-Tempered Aircraft." In the vote of thanks proposed by Society members immediately after the lecture, it was predicted that this paper would become a classic of technical literature.

Raymond set his theme with a bold analogy. "I propose," he said:

> to deal with that particular part of the "state of the art" that has to do with current principles and practices, rules and techniques, that have come to be generally associated with the design and production of well-conceived, well-executed, well-shaken down—that is, *well-tempered* aircraft.
>
> This title has more than a passing relationship to Bach's Well-Tempered Clavichord. The physical laws of musical harmony are essentially mathematical. Two tones whose vibrations are as 1 to 2, . . . or as 3 to 4, are more harmonious together than two whose vibrations are as 1 to 11 . . . but complete adherence to these mathematical laws, even when applied to a device as simple as the piano, would result in an instrument with a far greater number of tones than the piano and one incredibly difficult to play. The great achievement produced by tempering, or shifting slightly, the vibrations of individual tones is that, at an insignificant sacrifice in true harmony, there results a highly practical, highly flexible instrument well within the artistic capability of ordinary humans.

In his paper he drives through to demonstrate that the aircraft manufacturer must compromise between "true extremes" if he is to produce a "practical, highly flexible instrument." Raymond concludes his lecture with a telling anecdote about a technician who explained his approach by saying, "common sense is a rare gift of God. I have only a technical education."

Raymond's "Well-Tempered Aircraft" is itself a well-tempered paper, written by a well-tempered personality who will not, in the name of technical caution and objectivity, surrender the human right to over-all perspective. After reading Raymond, the technical writer or editor cannot again entertain, even in a total lapse of spirit, the notion that to be scientific, writing must be dull.

Conclusions

I remember once reading a book by a certain philosopher in which he considered, one per chapter, the major systems of philosophy. At the end he added a chapter called "Confession of Personal Faith." But it was entirely unnecessary to read it, and I never met anybody who had, because in the way he had been discussing systems other than his own, he had already betrayed his own beliefs. I am certain that my conclusions are by now no mystery to the reader, but for the sake of getting our discussion off the pad I shall restate them, at least the most important ones.

First of all, I hope that I have convinced some working writers and editors that there is some personal and professional value in studying the classics of technical literature. No matter how superior our age is, it cannot give us the best of everything. If I feel in the mood for music that resolves the chaos of life on a serene note, I have not betrayed my time if I put a Mozart disc on the hi-fi. Actually, Mozart can give me the strength to believe anew in the possibility of serenity even in my own time. And just as Raymond went back to Bach's great revision of the keyboard for his analogy for the well-tempered aircraft, so can we go back to writers of the past for some perspective on our present problems.

What, then, are some of the things that past technical writers can teach us? Beginning with smaller matters of organization and approach, I think they offer for our consideration a great variety of forms, especially in the way of handling definitions and descriptions of process, apparatus, and theory. This consideration of varied forms is important in a day when more and more technical writing is being turned out according to formula.

More important, I think the classics are reminders that even technical writing can be energetic, rhythmical, metaphoric without departing from scientific validity. This is important in a day when technical writing has lost so much of the art of connotation which in good writing must reinforce the science of denotation. Can there be any doubt that this energy, this rhythm, this metaphor, this imagery—in short, this *style*—are functions of the free personality, bringing its whole self to bear on the writing at hand?

But in the sacred temple of scientific objectivity we preach in effect, that technical writing must be grimly impersonal, that the personal pronoun and the adjective necessarily mean distortion and subjectivity, that the analogy is such a sure proof of fallacious approach, that one should not use it even for illustration. And so we encourage the split personality, we inhibit style, and give a professional haven to that bore who, as far as writing ability is concerned, is a perpetual parts-lister.

No one who has ever dipped into the classics of technical writing can have much patience with this approach. And anyone who swears by this impersonal approach has voluntarily converted himself into a timid, inhibited, colorless hack, and has resigned from the human race. I should say, *in my opinion* he has resigned from the human race. But didn't you know it was my opinion?

THE SCIENTIFIC METHOD AND FUNCTIONAL ENGLISH

Now that you have read an essay about the backgrounds of technical writing, two useful points of practical interest to technical writers can be made: Both the theoretical work of science and the practical work of technology depend directly upon the scientific method of inquiry, and clarity and conciseness as the chief aims of style in functional English are the natural outgrowth of the scientific method.

The scientific method has been several thousand years in developing. Progress in matters relating to the enlargement of the human mind usually occurs only gradually, at which point a Galileo or Faraday may appear to lead forward in a giant step. It is necessary for even the most obscure technologist to recognize that he is working, however quietly and unspectacularly, within a system or method which extends back to an Aristotle or forward to an Einstein. While it is true that scientists and engineers are preoccupied with what we may call theoretical work and that technologists are concerned generally with translating theory into action, all depend for their progress upon the scientific method, the chief steps of which may be summarized as follow:

1. Formulate a specific hypothesis or question for investigation.
2. Design an investigation to prove the hypothesis or to answer the question.
3. Accumulate data according to a design.
4. Classify the data.
5. Develop a generalization from the data.
6. Verify the results.

All technical writing may be referred to one or several steps in this process. It is obviously useful for the technician to know *where* what he is doing fits into the overall design.

CLARITY, PRECISION, AND OBJECTIVITY

By now it is evident that the style of functional technical English aims to be clear, precise, and scientifically objective. Yet experience in reading scientific and technical writing shows that much of it is unclear, imprecise, and subjective-sounding. What are some of the possible reasons for faulty style in technical writing?

1. *The technical writer may not know his subject thoroughly enough to write about it.* There are many reasons why this may be so, but usually the fault arises simply from writing without sufficient study and thought in advance. Perhaps the writer is merely the victim of the pressures of a difficult time schedule, but just as often he has not adequately appreciated the need for total mastery of his subject beforehand. The possibility of communicating clearly and precisely an imperfectly mastered subject is remote.

2. *While the writer may know the mechanical details surrounding his subject, he has not seen into the essence of the problem.* Such a situation is likely to produce a technical report of the "so what?" variety. The report is large on details but short on meaning. Substance is lacking from the form of the report. The writer has failed to lead his reader firmly through the abundance of minutiae to a clear and precise conclusion. The paper may be satisfactory in a mechanical way but fail to direct attention to the basics and essentials of the problem. Only the self-critically honest writer can overcome this fault, yet some technicians are unwilling to admit the difficulty even exists.

3. *The writer knows his subject very well, but he has not mastered the techniques of technical writing.* It is not enough to know well *what* one wishes to write about—one must understand *how* to transmit written information in accordance with the standard practices of technical writing. Private, personal technical writing techniques practiced only by one man are likely to confuse readers trained to understand a conventional standard. Public technical writing meant to be shared with others cannot easily be clear, precise, and objective if the

techniques it employs are not those accepted by the technical community.

4. *The technical writer has mastered both his subject and a correct style, but he does not know his audience.* Miscalculation of the level of understanding of one's readers easily leads to wasted effort. Technical writing is clear and precise merely in a relative way. Mastery of subject and style means little unless the accomplishments can be shaped to the needs of a particular audience. It is evident that only the technical writer with good control of both his subject and a standard writing style is capable of directing these to general levels of understanding of a particular audience.

Chapter 16
FUNCTIONAL ENGLISH (CONT.)

FUNCTIONAL SENTENCES

The technical writer must gradually gain sure control over his sentences. As the medium for expressing a complete idea, the sentence in English is the single most indispensable unit at the writer's command.

Studies have determined that the vast majority of sentences written in America today fall into one of five different structural patterns. These patterns are the fundamental molds or forms upon which all writers—but, especially, technical writers—depend for their work. Without an awareness of the skeletal frames of functional sentences, it is doubtful that any technician can achieve writing success.

FUNCTIONAL SENTENCE PATTERN 1

Subject and verb:

Technicians write.

The simplest communication names something (*technicians*) and says something (*write*) about it. When the something named (the subject) performs or carries out an action, the verb (*writes*) describing the action is active. The simple subject with active verb is the most basic and frequently used functional sentence pattern in English.

Chemicals mix.
Wheels turn.
Air flows.

FUNCTIONAL SENTENCE PATTERN 2

Subject, verb, and direct object.

Thermostats control heat.

In the second most frequently used sentence pattern, the object (*heat*) tells what or who directly receives the action performed by the subject (*thermostats*) of the verb (*control*). The subject (*thermostats*) names the performer of the action (*controls*) upon the direct object (*heat*). The reader knows who acts (as expressed by the verb) upon what.

Catalysts precipitate residue.
Oxidation produces vapor.
Pressure creates vacuum.

FUNCTIONAL SENTENCE PATTERN 3

Subject, verb, and predicate nominative.

Technicians are writers.

While in Pattern 2 the direct object indicates the receiver of the action named by the verb, in Pattern 3 the predicate nominative (*writers*) re-names or re-identifies the subject (*technicians*) by linking the subject and itself. Predicate nominative structure occurs with the verb *be* in any of its forms (am, is, are, was, were, being, been). *Be*, unlike other verbs, expresses a state of being rather than an action.

Slide rules are tools.
Ice was a hindrance.
Vibrators were dynamos.

FUNCTIONAL SENTENCE PATTERN 4

Subject, verb, and predicate adjective.

Coils were hot.

The predicate adjective structure also occurs with the verb *be* in any of its forms. Though the adjective (*hot*) follows the verb (or the *predicate*, were, as it is sometimes called), the adjective refers backwards to the subject (*coils*). Because the adjective appears to be with the verb, it is called the predicate adjective. But adjectives, by definition, identify or describe nouns or subjects. Hence, the true word modified must be found in the subject. The verb *be* is said to link its subject with either the predicate nominative or predicate adjective.

>*Writing is challenging.*
>*Air was contaminated.*
>*Calcium products were formed.*

FUNCTIONAL SENTENCE PATTERN 5

Expletive, verb, and subject.

>*There are three possible causes.*

In all four preceding functional sentence patterns the subject comes before the verb and is the first basic word of importance in the communication. The word order may be changed, however, by using an expletive (*there* or *it*) to take the grammatical place of the subject in front of the verb. The true subject (*causes*) remains grammatically unchanged, but it is withheld for greater emphasis until the end of the sentence. The expletive (*there*) simply stands for the real subject (*causes*) in order to achieve variety of sentence style as well as emphasis of meaning.

>*There are unexplained deviations.*
>*It is an unknown factor.*
>*There is leakage.*

Experts generally agree that no matter which sentence pattern a writer uses, the best sentences in basic technical writing are those which are shortest and most free of complicated wording. A beginning writer is sometimes tempted to display his own technical knowledge by unnecessarily using highly technical terms or long, involved sentences. Mastery of the five functional sentence patterns will help the writer eliminate superfluous words and phrases that add nothing to his essential meaning. In adapting words and sentences to his particular use, the technical writer should constantly ask:

What am I trying to say?
What basic, concrete, specific words will express it?
Which sentence pattern will make it clearest?
Can I say it more briefly and clearly?

STRUCTURED SENTENCES

Our discussion of sentences from the functional view has emphasized syntax—that is, the normal order of words in the commonly used sentence patterns. Structure of sentences may usefully be considered from a grammatical viewpoint as well. What are the various possibilities for combining the grammatical elements observed in the functional sentence patterns into diverse structures and forms?

A sentence is a group of words expressing a unit of thought and normally containing a subject and predicate. On the basis of grammatical form, sentences are classified as simple, compound, or complex.

Simple sentence. A simple sentence has one main clause. The subject or verb may be compound.

> *The mercury rises.*
> *Darwin, the great English zooligist, postulated a theory of the origin of the species.*
> *Ample supplies and personnel are needed.*
> *Lock the supply cabinet before leaving.*

If the technical writer forms a sentence according to the nature of the idea he wishes to express, it is clear that a simple sentence is best for stating an uncomplicated, unqualified observation.

Compound sentence. A compound sentence has two or more main clauses.

The initial flow of water contained traces of copper and sulfur, but these elements virtually disappeared as the pumping continued.

Release the lock; remove the tray; re-set the spring; start the timer.

There are several additional tests to carry out, and work is still proceeding on the first series.

If there are coordinate ideas to express in balance or contrast, the technician would be wise to adopt a compound sentence to communicate them effectively.

Complex sentence. A complex sentence has one main clause and one or more subordinate clauses.

When the final report comes from the laboratory, we will be able to make a more informed judgment.

A replacement for cast iron, which would cost less but function as well, has not been found.

The gears meshed poorly becaused they were not recently oiled.

The traffic coordinator asked when the shipment had been made.

The writer will find the complex sentence the most logical to use when there are several ideas to express in relationship to one another. The dependent clauses are used to express ideas subordinate to those in the main clause. The grammatical form of the sentence in itself helps express the nature of the relationship between the ideas—main and subordinate.

FUNCTIONAL PARAGRAPHS

Nowhere is the idea of functionalism in technical writing clearer than in paragraphing. The functional paragraph is adapted by form and structure to a special use.

Traditional methods of developing paragraphs by definition, classification, cause-effect, or process were treated earlier in this book. Paragraphs, regardless of the method used to form them, are structured around a single idea.

The topic sentence is the central thought around which a paragraph is built. In technical writing the topic sentence is usually the opening sentence of the paragraph and serves to tell the reader what function the unit will satisfy—what single segment of the report the writer is developing. The topic sentence may also appear as the concluding sentence of the paragraph. In this case it signals that the unit of thought is completed and clears the way for the start of the next unit.

In the following paragraphs the topic sentence is the initial sentence. The student may observe how the sentences serve to announce the particular function to which the paragraphs will be adapted.

SAMPLE 1

A common means of manufacturing acrylic polymers is through emulsion polymerization. The process combines the economy and safety of an aqueous reaction medium with a rapid but steadily controlled polymerization reaction. It provides an excellent yield of high molecular weight polymers. In emulsion polymerization, the following key ingredients are used: water, monomer, initiator, and emulsifier. The emulsifier plays an active role. It helps form and stabilize the acrylic polymer emulsion.

SAMPLE 2

Conventional instant coffee extract is sprayed into a drying tower where hot air dries the falling coffee droplets. The intense heat ($500°F.$) drives off some of the delicate aromatics and changes the coffee flavor. Rather than using heat, Product M is frozen at $40°$ below zero. Freezing locks in all the ground coffee flavor in Product M's superior blend.

SAMPLE 3

The dried prints are fired through a conveyorized furnace. A 45-minute cycle with a 6-7 minute peak at $760°$ is recommended. The optimum firing schedule, particularly rise rate and peak temperature, may vary with the type of furnace used. Peak firing temperatures of $725°-780°C$. can be used, but best results are generally obtained between $730°$ and $760°$. The furnace used should be capable of accurately maintaining the selected temperature profile. Peak temperature should be regulated to at least $\pm 2°C$.

The foregoing topic sentences indicate the function of the paragraphs. Information irrelevant or extraneous to the topic will not appear in the paragraph. The topic sentence may be thought of as introducing the subject of the paragraph or as summarizing the whole thought of the paragraph. While specific modes of developing paragraphs have been treated earlier in this text, logic usually demands that the topic sentence, once stated, must be developed in one of several ways:

1. Give facts or reasons which support it.
2. Give examples or specific details which explain or clarify it.
3. Give a sequence of events which illustrate it.

As the technician undertakes to develop a paragraph, he may ask himself:

1. What is the main point I wish to communicate in this paragraph?
2. What information must I provide to explain or clarify or support the main point?

When the paragraph is written, the writer may test its effectiveness by asking:

1. What is the central idea?
2. Has the reader been given enough information to make it clear?
3. Does the paragraph contain anything not related to the central idea?
4. Are the sentences of the paragraph in the most logical sequence of order to explain the topic sentence best?

a supplementary view...

THE SEVEN SINS OF TECHNICAL WRITING *By Morris Freedman*

From Morris Freedman, *"Seven Sins of Technical Writing,"* College Composition and Communication, *February 1958. Reprinted with the permission of the National Council of Teachers of English and Morris Freedman.*

Let me start by saying at once that I do not come to you tonight just as a professor of English, for, frankly, I do not think that I would have very much to say to you only as someone expert in the history of the use—and misuse—of the language. And any remarks on literature might be confusing, at least without extensive elaboration, for the values and objectives of literature seem so very different at first from those of technical writing—although fundamentally many of these values and objectives coincide. And I am sure that you are more than familiar with such things as cliches, comma splices, fragmentary sentences, and the other abominations we deal with in freshman composition. These obviously have nothing to do specifically with technical writing.

But I want to say, before anyone thinks that I class technical writing entirely by itself, immune from rules and requirements of communication that govern other kinds of writing, that technical writing calls for the same kind of attention and must be judged by the same standards as any other kind of writing; *indeed, it calls for a greater attention and for higher standards*. And I say this as a former science and medical writer for the popular press; as a former writer of procedure manuals and directives for the government; as a former editor of technical studies in sociology, statistics, law, and psychology; as a former general magazine editor; as a writer of fiction, essays, and scholarly articles; and, not least, as a professor of English. We can see at once why technical writing must be measured by higher standards, or, at least, by different ones, if anyone will not grant me that they are higher. Technical writing is so immediately functional. Confusing directions accompanying an essential device in a jet plane may result in disaster; bad writing elsewhere can have as its most extreme effect merely boredom.

Yet, while technical writing implicitly calls for great care, it differs from other kinds of writing in that its practitioners are, by and large, first technicians and only incidentally writers. And principally because of this arrangement, I think, technical writing has become characterized by a collection of sins peculiar to this discipline alone. I say the

253

collection is peculiar to technical writing, not any one of the sins alone. Any newspaper, weekly magazine, encyclopedia, textbook, any piece of writing you might name, will contain one or another of these sins, in greater or lesser profusion. But I know of no kind of writing that contains as many different sins in such great number as technical writing, and with such great potential for danger. To repeat, the sins in the world at large—at least, of the sort I'm talking about—often don't matter much. And sometimes, too, they don't matter in technical writing: "You got the meaning, didn't you?" Yes, I did, and so do we all get the meaning when a newspaper, a magazine, a set of directions stammers out its message. And I suppose, too, we could travel by ox-cart, or dress in burlap, or drive around with rattling fenders, and still get through a day.

But technical writing in this age can no more afford widespread sloppiness of expression, confusion of meaning, rattle-trap construction than a supersonic missile can afford to be made of the wrong materials, or be put together haphazardly with screws jutting out here and there, or have wiring circuits that may go off any way at all, or—have a self-destructive system that fails because of some fault along the way in construction. Technical writing today—as I need hardly reiterate to this audience—if it is much less than perfect in its streamlining and design may well result in machines that are less than trim, and in operation that is not exactly neat. This is at worst; at best, poor technical writing, when its effect is minimized by careful reading, hinders efficiency, wastes time. Let me remark too that the commission of any one of these sins, and of any one of many, many lesser ones, is really not likely alone to be fatal, just as one loose screw by itself is not likely to destroy a machine; but always, we know, sins come in bunches, the sin of avarice often links hands with the sin of gluttony, one loose screw many mean others, and, anyway, the ideal of no sins at all—especially in something like technical writing, where the pain of self-denial should be minimal—is always to be strived for.

A final word before I launch into the sins (whose parade, so long delayed, will prove, I hope, so much more edifying—like a medieval tableau). The seven I list might be described as cardinal ones, and as such they are broad and overlapping, perhaps, rather than specific and very clearly distinguished from one another. They all contribute to making technical writing less clear, concise, coherent, and correct than it should be.

Sin 1, then, might be described as that of *Indifference*, neglecting the reader. I do not mean anything so simple as writing down to an engineer or physicist, although this is all too common and may be considered part of this sin. This writing down—elaborating the obvious—is one reason the abstract or summary has become so indispensable a part of technical reports; very often, it is all the expert needs to read of the whole report, the rest being a matter of all too obvious detailing. Nor do I mean writing above the heads of your audience either, which is a defect likely to be taken care of by a thoughtful editor. Both writing over or under the heads of your reader, or to the side, are really matters of careless aiming and, as such, of indifference, too. But what I mean here by indifference are shortcuts of expression, elliptical diction, sloppy organization, bringing up points and letting them hang unresolved, improper or inadequate labelling of graphic material, and the like. This is communication by gutturals, grunts, shrugs, as though it were not worth the trouble to articulate carefully, as though the reader didn't matter—or didn't exist. This is basically an attitude of disrespect: *Caveat lector*—let the reader beware. Let the reader do his own work; the writer isn't going to help him.

Here is the concluding sentence from a quite respectable report, one most carefully edited and indeed presented as a model in a handbook for technical writers used by a great chemical firm. The sentence is relatively good, for it takes only a second reading to work out its meaning (perhaps only a slow first one for someone trained in reading this kind of writing):

When it is assumed that all of the cellulose is converted to ethyl cellulose, reaction conversion of cellulose to ethyl cellulose, per cent of cellulose reacted, and reaction yield of ethyl cellulose based on cellulose are each equal to 100%.

This is admittedly a tough sentence to get across simply, considering that "cellulose" is repeated in several different contexts. Yet two guiding principles would have made it much clearer: (1) always put for your reader first things first (here, the meaning hangs on the final phrase, "each equal to 100%," which comes at the end of a complicated series); and (2) clearly separate items in a series. (The second rule seems to me one of the most important in technical writing where so many things have to be listed so often.) Here is the recast sentence:

If all the cellulose is converted to ethyl cellulose, each of the following factors is then equal to 100%:
1. reaction conversion of cellulose to ethyl cellulose.
2. proportion of cellulose reacted.
3. reaction yield of ethyl cellulose based on cellulose.

The changes are not great, certainly, but in the process we have eliminated the indisputable notion of a percent being equal to a percent, and have arranged the series so that both the eye and the mind together can grasp the information immediately. Sin 1 then can be handled, one way, by cutting out indirect Rube Goldbergish contraptions and hitting your points directly on their heads, one, two, three.

The remaining sins I shall discuss are extensions of this primal one, disregard for the reader. Sin 2 may be designated as *Fuzziness*, that is, a general fuzziness of communication—vague words, meaningless words, wrong ones. The reader uses his own experience to supply the meaning in such writing; the writing itself acts only as a collection of clues. The military specializes in this sort of thing. I recall an eerie warning in an air force mess hall: "Anyone smoking in or around this mess hall will be dealt with accordingly." It still haunts me. Here is a caution in a handbook of technical writing with which you may be familiar: "Flowery, euphemistic protestations of gratitude are inappropriate." We know what this means, of course, but we ourselves supply the exact meaning. It happens that a "euphemism" is "the substitution of an inoffensive or mild expression for one that may offend or suggest something unpleasant." At least, that's what *Webster's Collegiate* says it is.

Here are some other examples: "The intrinsic labyrinth of wires must be first disentangled." The writer meant "network," not "labyrinth"; and I think he meant "internal" for "intrinsic" and "untangled" for "disentangled." Item: "The liquid contents of the container should then be disgorged via the spout by the operator." Translation: "The operator should then empty the container." Here is a final long one:

When the element numbered one is brought into tactual contact with the element numbered two, when the appropriate conditions of temperature have been met above the previously determined safety point, then there will be exhibited a tendency for the appropriate circuit to be closed and consequently to serve the purpose of activating an audible warning device.

Translation:

When the heat rises above the set safety point, element one touches element two, closing a circuit and setting off a bell.

Prescription to avoid Sin 2: use concrete, specific words and phrases whenever you can, and use only those words whose meaning you are sure of. (A dictionary, by the way, is only a partial help in determining the correct and *idiomatic* use of a word.) English is perhaps the richest of languages in offering a variety of alternatives for saying the same thing.

Sin 3 might be called the sin of *Emptiness*. It is the use of jargon and big words, pretentious ones, where perfectly appropriate and acceptable small and normal words are available. (There is nothing wrong with big words in themselves, provided they are the best ones for the job. A steam shovel is right for moving a boulder, ridiculous for picking up a handkerchief.) We may want to connect this sin with the larger, more univeral one of pride, the general desire to seem important and impressive. During World War II a high government official devoted much time to composing an effective warning for a sticker to be put above light switches. He emerged with "Illumination is required to be extinguished on these premises on the termination of daily activities," or something of the sort. He meant "Put the lights out when you go home."

The jargon I'm talking about is not the technical language you use normally and necessarily for efficient communication. I have in mind only the use of a big word or a jumble of words for something that can be said more efficiently with familiar words and straightforward expressions. I have in mind also a kind of code language used to show that you're an insider, somewhere or other: "Production-wise, that's a high-type machine that can be used to finalize procedure. The organization is enthused." There is rarely any functional justification for saying "utilize" or "utilization" for "use," "prior to" for "before," "the answer is in the affirmative or negative" for "yes or no," or for using any of the "operators, or false verbal limbs," as George Orwell called them, like "render inoperative," "prove unacceptable," "exhibit a tendency to," "serve the purpose of," and so on and on.

Again, one can handle this sin simply by overcoming a reluctance to saying things directly; the most complex things in the world can be said in simple words, often of one syllable. Consider propositions in higher math or logic, the Supreme Court decisions of men like Brandeis and Holmes, the poetry of Shakespeare. I cannot resist quoting here Sir Arthur Quiller-Couch's rendition in jargon of Hamlet's "To be or not to be, that is the question." I am sure you all know the full jargon rendition of the soliloquy. "To be, or the contary? Whether the former or the latter be preferable would seem to admit of some difference of opinion."

Sin 4 is an extension of 3: just plain *Wordiness*. The principle here is that if you can say anything with more words than necessary for the job, then by all means do so. I've already cited examples of this sin above, but compounded with other sins. Here is a purer example, the opening of a sentence in a technical writing handbook: "Material to be contained on the cover of the technical report includes . . ." This can be reduced to "The cover of the technical report should include . . ." Another example, less pure: "The front-mounted blade of the bull-dozer is employed for earth moving operations on road construction jobs." Translation: "The bull-dozer's front blade moves earth in road building." Item: "There is another way of accomplishing this purpose, and that is by evaporation." Translation: "Evaporation is another way of doing this." Instead of saying simply that "the bull-dozer's front blade moves earth," you say it "is employed for earth moving operations," throwing in "employed" and "operations," as though "moves" alone is too weak to do this tremendous job. The cure for this sin? Simply reverse the mechanism: say what you have to in the fewest words.

Sin 5, once again an extension of the immediately preceding sin, is a matter of *Bad Habits*, the use of pat phrases, awkward expressions, confusing sentence structure, that have, unfortunately, become second nature. Again, I'm not alluding to the perfectly natural use of familiar technical expressions, which may literally be called clichés, but which are not efficiently replaceable. Sin 5 is a matter of just not paying attention to what you say, with the result that when you do suddenly pay attention, you see the pointlessness or even humor of what you have set down. Perhaps the most common example of this sin is what has been called "deadwood," or what may be called "writing for the simple minded." Examples: "red in color," "three in number," "square in shape," "the month of January," "the year 1956," "ten miles in distance," and the like. What else is red but a color, three but a number, square but a shape, January but a month, 1956 but a year, ten miles but a distance? To say that something is "two inches wide and three inches long" is to assume that your reader can't figure out length and width from the simple dimensions "two inches by three inches." I once read that a certain machine was 18 feet high, "vertically," the writer made sure to add; and another time that a certain knob should be turned "right, in direction."

A caution is needed here. There are many obvious instances when qualification is necessary. To say that something is "light," for example, is plainly mysterious unless you add "in color" or "in weight" or, perhaps, "in density" (unless the context makes such addition "deadwood").

I would include under Sin 5 the locutions "as far as that is concerned" (lately shortened to "as far as that"), "as regards," "with regard to," "in the case of" ("In the case of the case enclosing the instrument, the case is being studied"). These are all too often just lazy ways of making transitions (and, thus, incidentally, quite justifiable when speed of writing is a factor).

Sin 6 is the *Deadly Passive*, or, better, deadening passive; it takes the life out of writing, making everything impersonal, eternal, remote and dead. The deadly passive is guaranteed to make any reading matter more difficult to understand, to get through, and to retain. Textbook writers in certain fields have long ago learned to use the deadly passive

FUNCTIONAL ENGLISH (CONT.)

to create difficulties where none exist; this makes their subject seem weightier, and their accomplishment more impressive. (And, of course, if this is ever what you have in mind on an assignment, then by all means use the deadly passive.) Sin 6 is rarely found alone; it is almost indispensable for fully carrying out the sins of wordiness and jargon. Frequently, of course, the passive is not a sin and not deadly, for there simply is no active agent and the material must be put impersonally.

Examples of this sin are so easy to come by, it is difficult to find one better than another. Here is a relatively mild example of Sin 6.

The standardization of procedure in print finishing can be a very important factor in the efficient production of service pictures. In so far as possible, the smallest number of types and sizes of paper should be employed, and the recommended processing followed. The fewer paper grades and processing procedures used, the fewer errors and make-overs that are likely. Make-overs are time-consuming and costly.

Here it is with the deadly passive out and some other changes made:

To produce service pictures efficiently, a standard way of finishing prints can be very important. You should use as few types and sizes of paper as possible, and you should follow the recommended procedure for processing. In this way, you will make fewer errors, and have to re-do less work. You will save time and money.

Associated with the deadly passive, as you might see from the two passages above, is the use of abstract nouns and adjectives for verbs. Verbs always live; nouns and adjectives just sit there, and abstract nouns aren't even there. Of course, there are a number of other ways of undoing the passivity of the passage I quoted, and of making other improvements, just as there were other ways of handling any of the specimens I have cited in the train of horrors accompanying my pageant of sins.

Finally we come to Sin 7, the one considered the deadliest by many, and not only by teachers of English but by technical writers and technologists of various sorts: *Mechanical Errors*. I don't think this sin the deadliest of all. It does happen to be the easiest one to recognize, the one easiest to deal with "quantitatively," so to speak, and the easiest one to resist. I suppose it is considered deadliest because then those who avoid it can so quickly feel virtuous. It can promptly be handled by good works alone. Actually most technical writing happens to be mechanically impeccable; not one of the examples I have used tonight had very much mechanically wrong with it. If anything, technical people tend to make too much of formal mechanics. I remember working with a physicist who had much trouble saying anything in writing. While his general incapacity to write was almost total, one thing he did know, and know firmly, and that was that a split infinitive was to be abhorred. That, and using a preposition to end a sentence with. He could never communicate the simplest notion coherently, but he never split an infinitive or left a preposition at the end of a sentence. If Nobel Prizes were to be awarded for never splitting infinitives or for encapsulating prepositions within sentences, he would be a leading candidate.

There are a handful of mechanical errors which are relevant to technical writing, and these are important because they are so common, especially in combination with other sins. (Split infinitives or sentence-ending prepositions, need I say, are not among them.) These are dangling participles and other types of poorly placed modifiers, and ambiguous references. There are others, a good number of others, but the ones I mention creep in most insidiously and most often.

Here are some examples stripped down to emphasize the errors:

Raising the temperature, the thermostat failed to function.
Who or what raised the temperature? Not the thermostat, I presume; and if it did somehow, as the result of current flowing in its wiring, then this ought to be said quite plainly.

The apparatus is inappropriately situated in the corner since it is too small.
What is too small? Apparatus or corner?

Every element in the device must not be considered to be subject to abnormal stress.
What is meant here is that "Not every element in the apparatus must be considered subject to abnormal stress," almost the opposite of the original taken literally.

I should like to conclude by emphasizing something I glanced at in my introduction, that the seven sins of technical writing are to be avoided not so much by a specific awareness of each, accompanied by specific penance for each, as by a much more general awareness, by an attitude toward subject matter, writing process, and reader that can best be described only as "respectful." You will not help yourself very much if you rely on such purely mechanical aids as Rudolf Flesch's formulas for "readable writing," or on slide rules measuring readability, much as you may be tempted to do so. These can be devil's snares, ways to make you think you are avoiding sin. There are no general texts, either, at present that will help you in more than very minor ways. The only aids you can safely depend on are the good book itself, that is, a good dictionary (there are many poor ones), any of the several volumes by H. W. Fowler, and occasional essays, here and there, by George Orwell, Jacques Barzun, Herbert Read, Somerset Maugham, and others. And these, I stress, can only be *aids*. What is most important in eliminating sin in technical writing is general attitude—as it may well be in eliminating sin anywhere.

I repeat that technical writing must be as rationally shaped as a technical object. A piece of technical writing, after all, is something that is shaped into being for a special purpose, much as a technical object. The design engineer should be guided in his work by the requirements of function almost alone. (Of course, if he happens to have a boss who likes to embellish the object with useless doo-dads, why then he may have to modify his work accordingly to keep his job—as automobile designers do every day; but we try never to have in mind unreasonable situations of this sort.) It is as pointless for the design engineer to use three bolts where one would do (both for safety and function), to make an object square when its use dictates it should be round, to take the long way through a process when there is a short way, as it is for the technical writer to commit any of the sins I have mentioned. Technical writing—informative writing of any sort—should be as clean, as functional, as inevitable as any modern machine designed to do a job well. If I will not be misunderstood in throwing out this thought, I should like to suggest to you that good technical writing should be like good poetry—every word in its exact place for maximum effect, no word readily replaceable by another, not a word too many or too few, and the whole combination, so to speak, invisible, not calling attention to its structure, seemingly effortless, perfectly adapted to its subject.

If one takes this general approach to the shaping of a piece of technical writing, and there really can't be much excuse for any other, then there is no need to worry about any of the sins I mention. Virtue may not come at once or automatically, for good writing never comes without effort, however fine one's intentions, but it will certainly come, and perhaps even bring with it that same satisfaction the creative engineer experiences. Technical writing cleansed of its sins is no less worthy, no less impressive, an enterprise than good engineering itself. Like mathematics to physics, technical writing is a handmaid to technology, but like mathematics, too, it can be a helpmate, that is, an equal partner. But it can achieve this reward of virtue only by emphasizing the virtues of writing equally with those of technology.

Chapter 17
FUNCTIONAL ENGLISH (CONT.)

FUNCTIONAL GRAMMAR

By definition, functional grammar is correct grammar. Incorrect grammar prevents understanding and is therefore the enemy of technical writing. The belief that grammar is not important to the technician is erroneous. Grammatically correct technical writing is familiar and generally accepted—it is easier to read and understand than ungrammatical writing. Public writing *must* be grammatical; private writing *may* be ungrammatical. The technician who argues against this is probably making illogical excuses for his own deficiencies.

This is a textbook on technical writing; it is not a grammar, a rhetoric, or a book devised for the general teaching of college English. It deals, therefore only with those aspects of usage and grammar which are necessary to satisfy its primary mission. The survey of grammatical problems which follows is brief and highly selective, with the needs of the writer-technologist in view. This review of aspects of functional grammar—together with the Guide to Good Use and the Guide to Grammatical Terms—will be sufficient for the majority of basic technical writers. Others may require a review of grammar outside the scope of the present writing.

The Question of Number

A. A verb agrees in number with its subject.

Improved combustion engines *are* our greatest need.
Our greatest need *is* improved combustion engines.

B. A pronoun or pronominal adjective agrees in number with its antecedent.

The two outside brackets are B-37 and B-38 on the diagram. *They* support the circuit-housing.
Circuit-housing is supported by outside brackets B-37 and B-38 on the diagram. *It* cannot be properly positioned before the brackets are in place.

C. A compound subject coordinated by *and* requires a plural verb, regardless of the individual number of the member subjects.

Detergents and the wire brush *are* kept beneath the counter.

D. When the compound subject refers to a single unit, a singular verb is used.

Boyd & Fraser *is* a San Francisco publishing company.
The block and tackle *is* used to hoist heavy equipment.

E. The number of the verb is determined solely by the number of its subject. Nouns and pronouns appearing between a subject and verb have no effect upon the number of the verb.

The textile expert, together with several helpers, *is* working out the dye formula.
The design engineers, as well as the controller, *are* meeting with research management.

F. Collective nouns may be either singular or plural, depending on whether the whole or the individual membership is emphasized.

The Research Executive Committee *establishes* all work priorities.

Members of the Research Executive Committee *establish* priority of the work.

G. Expressions of aggregate quantity, even though plural in form, generally are construed as singular.

Three-fourth's of our effort *is* spent in preparation.
A thousand r.p.m.'s *is* the usual speed.
Two times this year's growth *is* anticipated for next year.

H. When expressions of quantity do not merely specify an aggregate amount, but stress the units that compose the aggregate, they are construed as plural.

Three hundred units *were* purchased.
Thirty bags of cement *are* stored in the shed.

I. A relative pronoun should not be taken as singular when its antecedent is a plural object of the preposition *of* following the word *one*.

F-249 is one of the most potent chemicals which *are* available to do *their* work in metallurgy.
Simpson is one of the experts who *have* been called in to offer *their* opinions.

The Question of Antecedents

A. Pronouns must have definite antecedents or antecedents clearly understood.

The elevator was overloaded and could not carry it. (The antecedent here is merely implied but not expressed.)
The overload was so great that the elevator could not carry it. (Revised to make the antecedent clear.)
The elevator could not carry the overload. (Revised to eliminate the troublesome pronoun.)

B. Vague second-person reference of pronouns should be avoided.

Ample time for preliminary preparation must be allowed. You

must plan well ahead. (Unsatisfactory, vague, second-person *you*.)

Ample time for preliminary preparation must be allowed. The technician must plan well ahead. (Revised to include a common noun in place of vague pronoun.)

C. Appositives must have clear antecedents. Vague appositive reference can be corrected by placing the appositive immediately after the word or phrase it modifies or by rewording the sentence.

The chief technologist devised a new system for wiring the amplifier, while experimenting with the circuit, a greatly improved method of procedure. (Vague and unclear reference to antecedent.)

While experimenting with the circuit, the chief technologist devised a new system for wiring the amplifier, a greatly improved method of procedure. (Revised.)

The chief technologist greatly improved the procedure by devising a new system for wiring the amplifier. He developed the new procedure while experimenting with the circuit. (Revised.)

The Question of Dangling Modifiers

Expert observers generally agree that the most prevalent grammatical fault in technical and scientific writing is the dangling modifier. Such modifiers are usually verbals which have no word to which they can logically attach themselves and are thus said to dangle.

After filling the tank, the piston broke down. (Dangling.)
After the tank was filled, the piston broke down. (Revised.)
When shifting gears, oil pressure is at the maximum. (Dangling.)
Oil pressure is at the maximum during the shifting of gears. (Revised.)
Hovering over the kiln, the ceramist observed a slightly malodorous gas. (Dangling.)
The ceramist observed a slightly malodorous gas hovering over the kiln. (Revised.)

The Question of Misplaced Modifiers

A. Modifiers must be placed near the words they modify.

Place the nut outside the shield in the tooled out housing aperture. (Inaccurate and confusing.)
Place the nut in the tooled out housing aperture outside the shield. (Revised to clarify intended meaning.)

B. Proper placement of *only* as a modifier is important in technical writing. Meanings may be altered substantially by shifting *only*.

The technician *only* altered the formula for dye in the process.
Only the technician altered the formula for dye in the process.
The technician altered *only* the formula for dye in the process.
The technician altered the formula for *only* the dye in the process.
The technician altered the formula for dye in the *only* process.
The technician altered the formula for dye *only* in the process.
The technician altered the formula for *only* dye in the process.

The Question of Illogical Shifts

A. Avoid illogical changes in the tense of verbs.

Acrylic polymer emulsions are employed in diverse technological applications. The emulsifier *played* an active role. It helps form and stabilize the acrylic polymer emulsion.
Acrylic polymer emulsions are employed in diverse technological applications. The emulsifier *plays* an active role. It helps form and stabilize the acrylic polymer emulsion. (Revised.)

B. Avoid illogical changes in the pronominal person.

Apply a thin coating of the laminate to the surface. You can remove any excess with a clean cloth. Bring the two sealing surfaces together.
Apply a thin coating of the laminate to the surface. Remove any excess with a clean cloth. Bring the two sealing surfaces together. (Revised.)

C. Avoid unnecessary changes in the voice of the verb.

A time-study engineer performed the analysis, and the statistics were soon tabulated.
A time-study engineer performed the analysis and soon tabulated the statistics. (Revised.)

D. Avoid irrational changes in the mood of verbs.

Work proceeds best with planning and methodology. Do not hurry the testing, but the schedule should be honored.
Work proceeds best with planning and methodology. The testing remains unhurried, and the schedule is honored. (Revised.)

E. Avoid unreasonably shifting sentence patterns.

It was due to careful study that made us select the successful system.
Careful study helped us select the successful system. (Revised.)
Due to careful study, we were able to select the successful system. (Revised.)

F. Avoid the shifted construction which results from failure to maintain grammatical parallelism. Parallelism requires use of similar grammatical structure in writing clauses, phrases, or words expressing ideas or facts which are equal in value.

Indoctrination procedures well planned and which are carried out carefully result in good training of personnel.
Indoctrination procedures well planned and carefully carried out result in good training of personnel. (Revised.)
These are chemical technologists with good experience and familiar with standard operating procedures.
These are chemical technologists with good experience and a knowledge of standard operating procedures. (Revised.)
Before engaging the circuit, the electrician should both verify the heat reading and he should be certain the ventilating device is opened.
Before engaging the circuit, the electrician should both verify

the heat reading and check to see that the ventilating device is opened. (Revised.)

The assembly line extends from the cutting machine, runs across the stamping operation, and into molding area.

The assembly line extends from the cutting machine to the stamping operation into the molding area. (Revised.)

Proper Use of the Passive Voice

Because the passive voice of the verb is more consistent with impersonal, objective writing than the active voice, it is more widely used in technical and scientific writing than anywhere else. Technical writing is more concerned with *work done* than with the person or persons doing it. The passive voice, accordingly, throws emphasis on the subject of the sentence—the subject of the verb is acted upon. (The active voice indicates the subject as performing the action, which stresses the performer of the action rather than the action itself.) Moreover, the passive voice functions more naturally and effectively than the active with the third person impersonal point of view, which is a chief characteristic of technical writing.

In the example which follows, the sentences are presented first in the traditional third person passive voice. Then, the same sentences are offered in the first person active voice. The greater efficiency and objectivity of the passive over the active voice can be readily observed:

To improve storage and freeze-thaw stability and to increase the resistance of the emulsion product to mechanical stress, a surfactant was post-added to the emulsion. The surfactant included the same emulsifiers that were used in the emulsion polymerization process. (Preferred passive voice.)

To improve storage and freeze-thaw stability and to increase the resistance of the emulsion product to mechanical stress, I post-added a surfactant to the emulsion. I selected a surfactant that included the same emulsifiers I used in the emulsion polymerization process. (Inappropriate active voice.)

There are, nonetheless, in technical writing many occasions when use of the passive voice is both unwarranted and unnecessary. More often than not, these occasions arise in technical memoranda and letters. In the group of examples which follow, comparison of the passive with the active voice shows that the former may be wordy, roundabout, and confusing on occasions when there is small justification for using it.

It is the desire of the Quality Control Engineer to determine how the error occurred in failing to add the storage-stabilizing materials during the process of manufacture. (Poor use of passive voice.)

Quality Control (or I or We) wants to know how the error occurred. Why were the storage-stabilizing materials not added during the process of manufacture? (Revised.)

This precaution may be justified in the interest of safety. (Poor use of passive voice.)

Safety may justify this precaution. (Revised.)

It is believed that the government should increase the prime rates of interest. (Poor use of passive voice.)

We believe the government should increase the prime rates of interest. (Revised.)

Avoiding Verbosity and Tautology

Verbosity means "wordiness," and tautology means "needless repetition in other words." These are two diseases of technical writing which may prove fatal in our age of mimeography machines and typewriters. All of us may be submerged and drowned in our own sea of words. The technical writer should try to save himself and others from this fate by avoiding padding and excess word-baggage.

Verbose: On the basis of the foregoing discussion and preceeding results, it can be said that....
Revised: The discussion and results thus far show that....

Verbose: Prior to the beginning of the investigation....
Revised: Before the investigation began....

Verbose: All other things being equal, the field workers are prepared to enter into their prescribed activities.

Revised: Field workers are ready to begin their planned activity.

Verbose: It is often desirable to add acrylic emulsions directly to a system without pre-neutralization since this will eliminate the handling of a highly viscous solution entirely.

Revised: Often the acrylic emulsions may be added directly to a system without pre-neutralization, thus eliminating entirely the handling of a highly viscous solution.

Part IV
THE TECHNICIANS' GUIDE TO GOOD USE OF ENGLISH

THE TECHNICIANS' GUIDE TO GOOD USE OF ENGLISH

Without any attention whatever to those matters of good usage cited in the following *Guide*, one may write adequate technical reports. That is true because standards of technical writing in some sectors of industry are not very demanding. Persons called upon to write the reports pride themselves on their technical proficiency but are indifferent to their writing practices.

The availability of persons capable of writing well about technology has not kept pace with the growth of technology. Unless the number of effective technical writers increases, however, it is likely that continued growth of technology will be impeded. Technology and written communication are closely allied—information must be communicated clearly and rapidly if progress is to be unabated.

The aspiring technical writer can bring to his effort a high degree of professionalism by developing at the start a sense of good English usage. This *Guide* is admittedly incomplete and abbreviated, even though it attempts to list a number of the most common problems of use. Yet the writer who masters these materials will have moved far toward upgrading the quality of his writing and developing a proper respect for preferred English usage.

A, An. Use *a* before a consonant sound, and *an* before a vowel sound or a silent *h*:

a ditch, a current, a watt, a heat
an accelerator, an engine, an industry
an ohm, an uptake, an hour

Above. *Above* is properly a preposition (above the fire) or an adverb (heat flowed above), but it is acceptable in technical writing when used as an adjective or noun:

(*Adjective*). Observe the *above* diagram.
(*Noun*). The *above* provides proof.

Absolutely. *Absolutely* is correct in formal and technical English in the sense of "entirely" or "completely," but not as a substitute for "very" or "very much":

Statistical analyses proved *absolutely* correct.
The research team was very (not *absolutely*) disheartened by the reduced budget.

Accept, Except. These verbs are similar in sound but different in meaning. *Accept* means "to receive" or "to give an affirmative answer to." *Except*, infrequently used as a verb, means "to exclude." As a preposition, *except* means "with the exclusion of":

The nut conveniently *accepts* the bolt.
Industry *accepted* the new standards.
Sodium formulations were *excepted* from the changes.
All formulations *except* those with sodium were changed.

Adapt, Adopt. To *adapt* means "to change something for a purpose." To *adopt* means "to take possession of":

The production schedule was altered to *adapt* to the new packaging procedures.
Quality-control personnel will *adopt* a different method of testing for impurities.

Advice, Advise. *Advice* is a noun and *advise* is a verb. They are pronounced, spelled, and used differently:

The electrical workers sought *advice* from the engineering unit.
The electrical workers agreed to do what the engineers might *advise*.

Advice. See *Advice*.

Affect, Effect. These verbs are similar in sound but different in meaning. *Affect* means "to influence," and *effect* means "to bring about." As a noun, *effect* means "result":

Cold air *affects* the drying time.
Heat *effects* a change in the viscosity of the mixture.
Change of altitude produced an *effect* upon the readings.

Aggravate. In technical and in formal English, *aggravate* means "to intensify." The term may not be used in technical writing with its colloquial meaning of "to annoy" or "to provoke":

Itching skin *aggravated* the discomfort.
Friction between the materials *aggravated* the difficulty of welding.

Agree to, Agree with. It is correct to *agree to* something and to *agree with* someone:

The marine scientists *agreed to* the design of the investigation.
We *agree with* the project manager that greater precautions are needed.

Alibi. In formal and technical English *alibi* means "a plea of having been elsewhere than at the alleged place where an act was committed." The term may not be used in technical writing with its colloquial meaning of "an excuse":

(*Right*) There was an *explanation* for the failure.
(*Wrong*) There was an *alibi* for the failure.

All the farther, All the faster. Formal and technical writing requires the use of *as far as* and *as fast as*:

As far as the test ran, no defects appeared in the material.
The first trial tested the blade *as fast as* 1500 r.p.m.'s.

Allude, Refer. To *allude* is to refer to something indirectly; to *refer* is to mention something specifically:

If the report *alludes* to difficulties in production, it is *referring* to the breakdown of the heavy die cutting machine.

Among, Between. *Among* implies more than two persons or things, while *between* implies only two. *Between* may be used when three or more things are considered individually:

Correlations were made *between* batches 3 and 7.
Comparisons were made *between* stations 11, 17, and 20.

Amount, Number. *Amount* refers to quantity or mass, while *number* refers to objects countable:

Each experiment requires the same *amount* of material.
The *number* of turns given was carefully calibrated to the operating mechanism.

Angle. In technical writing *angle* may be used only in its formal or mathematical meaning, not in its slang sense of "point of view" or "aspect":

(*Yes*) The draftsman determined the *angle* of the intersecting vectors.
(*No*) The study developed a new *angle* about weightlessness.

Anyone, Any one. *Anyone* means "any person at all." *Any one* singles out one person or thing in a group:

Anyone is capable of doing the job.
Any one of the new plugs may be used.

Allusion, Illusion. These nouns are similar in sound but different in meaning. *Allusion* means "an indirect reference." *Illusion* means "a misleading image" or "false impression":

The report made an *allusion* to recent studies carried out in the Soviet Union.
Diffusion of light created an *illusion* of breakage.

Already, All ready. *Already*, an adverb, means "previously." *All ready*, an adjective phrase, means "completely prepared":

Phase 1 was *already* completed.
The third batch must be *all ready* to receive the fresh plankton.

Also. *Also* should not be used as a substitute term for the conjunction *and*:

The catalyst precipitated out several tiny crystals *and* (not *also*) some minute particles.

Altogether, All together. *Altogether* is an adverb that means "wholly, completely." *All together* is an adjective phrase that means "in a group":

Results were *altogether* satisfactory.
Statistical analyses were computed *all together* for the final sum.

A.M., P.M. *a.m.* and *p.m.* are properly used only when preceded by a number:

The work will be completed by 3 p.m. (not *by* or *in the p.m.*).

Apt, Likely. In formal and technical writing *apt* refers to a natural ability or habitual tendency. *Likely* refers to a probability:

The surveyor is *apt* at measuring distance.
The mixture is *likely* to evaporate unless temperature controls are used.

A piece, Apiece. *Piece* is a noun, while *apiece* is an adverb:

A piece of debris contaminated the solution.
The three white rats were given a gram of the drug *apiece*.

As a method of. Often an overused phrase in technical writing when followed by a gerund:

(*Wordy*) Screening is useful *as a method of* separating components.
(*Better*) Screening is useful *for separating* components.

At. *At* should not ever be used after forms of *to be* in expressing the idea of location. This use is redundant and illiterate:

We knew where we were (not *were at*) in our research.

At about, At around. Doubling these prepositions is unnecessary and produces inexactness in technical writing. Either *at, about,* or *around* should be used:

(*Inexact*) Stop turning *at about* 30 degrees.
(*Exact*) Stop turning *at* 30.08 degrees.

Awhile, A While. *Awhile* is an adverb. *A while* consists of an article and a noun:

Accelerate *awhile* during the last phase.
Accelerate for *a while* during the last phase.

Bad, Badly. *Badly* is correct in formal and technical English only as an adverb modifying another adverb or an adjective. *Bad* is correct in all other instances:

The shipment contained several *badly* damaged articles.
There was a *bad* odor in the laboratory.

Balance. In technical writing *balance* should not be used in its colloquial sense of "the rest" or "remainder":

(*Poor*) The *balance* of the experiment was not completed.
(*Good*) The *remainder* (or *rest*) of the experiment was not completed.

Being that, Being as how. These are awkward and illogical substitutes for the subordinating conjunctions *as, because, since*:

(*Awkward*) *Being that* the animals were still under sedation, we waited 24 hours to continue the observations.
(*Better*) *Since* the animals were still under sedation, we waited 24 hours to continue the observations.

Beside, Besides. These are interchangeable prepositions in the sense of "except." Otherwise, *beside* means "by the side of" and *besides* means "in addition to":

Use the mortar *beside* the pestle.
Use the mortar *besides* the coarse grinder.

Bunch. In formal and technical English *bunch* should not be used to mean "several" or "a group":

There were *several* (not a *bunch of*) lumps of discolored material near the bottom of the tube.
A *group* (not a *bunch*) of test-rabbits will serve as placebos.

Bursted, Bust, Busted. None of these archaic or slang forms is accepted in formal or technical writing. The principal parts of the verbs are *burst, burst, burst*:

The cylinder was *burst* at the seam.
Mercury *burst* through the glass.

Can, May. In formal and technical writing *can* means "to be able," and *may* means "to have permission." *May* is correctly used in the sense of possibility:

The technician *can* start by dismantling the rear bracket, but he *may* find his access blocked by separated housing.

Cause, Reason. *Cause* is "that which effects a result," while *reason* is "an explanation of an act" or of a cause:

The *cause* of the collapsed girder was not yet explained.
Engineers were examining several possible *reasons* for the failure.

Censure, Criticize. To *censure* is to condemn as wrong; to *criticize* is to examine and judge:

Careless workmanship merits *censure*.
The willingness to *criticize* was taken as a sign of interest in the project.

Cite, Sight, Site. *Cite* is a verb meaning to refer to directly or to quote. *Sight* may be a verb or a noun to signify "a view" or "the ability to see." *Site* is a noun meaning "a location":

Cite the source of your information in a footnote.
It was difficult to *sight* the flaw in the surface because of the glare.
They are seeking a *site* for a new factory.

Climactic, Climatic, *Climactic* refers to climax; *climatic* refers to climate:

The *climactic* stage of cellular division occurred after 72 hours.
Climatic changes hindered the growth of the outdoor specimens.

Claim. Formal and technical English uses *claim* in its correct sense of "to demand as one's right":

(*Poor*) Smith *claims* there is no alternative method.
(*Better*) Smith *maintains* there is no alternative method.
(*Correct*) Jones *claimed* the prize he had won in biology.

Common, Mutual. *Common* has to do with things equally shared, while *mutual* refers to reciprocity:

Three of the schools pooled their resources and purchased the equipment for *common* use.
A *mutual* exchange of energy was sustained between the two batteries.

Complement, Compliment. *Complement* is that which fills up or completes. *Compliment* is an expression of praise:

The work of each of the two technicians *complements* that of the other.
Health of the children was the best *compliment* the dental hygienists could receive.

Considerable. In formal and technical English *considerable* should be used only as an adjective and not as a noun or adverb:

A *considerable* amount of residue remained.

Continual, Continuous. *Continual* means "frequently repeated." *Continuous* means "without interruption":

There were *continual* waves of sound emitted in decreasing volume.
Continuous pressure must be maintained in the test chamber.

Council, Counsel. A *council* is an assembly summoned for consultation. *Counsel* as a noun means *advice* and as a verb *to give advice*:

A *council* of scientists was directing the research.
The laboratory technician sought *counsel* from the professor on the correct procedure.

Criteria. See *Data*.

Data, Criteria, Phenomena. These are plural forms of *datum*, *criterion*, and *phenomenon*. *Data* may be used as a collective noun:

This single *criterion* was applied to all the *data*.
These *criteria* were measured against the *phenomenon* of light.
All *data* were subjected to statistical analysis.

Deal. In the sense of "transaction," *deal* should not be overused in favor of more exact terms such as *sale, agreement,* or *plan*:

An *agreement* was made with an outside testing laboratory.

Device, Devise. A *device* is "that which is devised or formed by stratagem," for example a tool for performing a particular operation. *Devise* means to form or to invent:

A special *device* was developed to permit separation of the components while the machine is in operation.
Electrical engineers *devised* the special switch.

Differ from, Differ with. *Differ from* means "to stand apart because of unlikeness." *Differ with* means "to disagree":

The two specimens *differ from* one another in several details.
Reports from two sources *differ with* one another on this point.

Dissimulate. See *Stimulate*.

Due to. *Due* is acceptable now both as an adjective and a preposition introducing an adverbial phrase. But *because of* and *owing to* are still excellent substitutes:

(*Adjective*) Success was *due to* continual effort.
(*Preposition*) Work was halted *due to* a strike.
(*Alternative*) Work was halted *because of* (or *owing to*) a strike.

Each other, One another. In formal and technical English, *each other* refers to two people only; *one another* refers to more than two:

The two government experts agreed with *each other*. The members of the panel could not completely agree with *one another*.

Effect. See *Affect*.

Either, Neither. Properly used to refer to one or the other of two, the words are singular subjects:

Either batch 27 or 34 is contaminated. *Neither* of the batches is yet cooled.

Environment. *Environment* still means "the aggregate of all the external conditions and influences affecting the life and development of an organism." It is wrong in formal and technical English to use the word to mean "neighborhood," "atmosphere," or "surroundings":

Ecology deals with the mutual relations between organisms and their *environment*.

Everyone, Every one. The pronoun *everyone* includes all the persons or things as a group. *Every one* refers to persons or things as separate entities:

Every one has his particular view of the need for procedural changes in the experiment.
Every one of the test tubes was thoroughly sterilized.
Everyone in the club planned to attend the next meeting.

Except for the fact that. This is a common but unnecessarily wordy substitute for *except that*.

Factor. *Factor* should properly be used to mean one of several elements contributing to a result, and not in the sense of a general feature or aspect:

A general *feature* (not *factor*) of the disease is metabolic in nature.
Accelerated heartbeat was one of the *factors* suggesting thyroidal imbalance.

Farther, Further. *Farther* appears to be the preferred word to express geographic distance. *Further* is used to express additional time, degree, or quantity:

Die cutting machine 3 is *farther* from the leading ramp than machine 2.
Further study of the effect of the pill will be needed.

Fewer, Less. In formal and technical English *fewer* refers to number while *less* refers to degree or quantity:

Fewer than a third of the test animals survived. There was *less* evaporation in the second culture medium.

Finalize. Some expert technical writers still object to the use of this form, associating its use with poorly written government and official documents:

All preliminary preparations *will be made final* (not *finalized*) by Thursday.

Fix. Do not use *fix* in formal and technical writing in the sense of arrange, prepare, or repair:

The packaging equipment must be repaired (not *fixed*) before we can resume production.

Formally, Formerly. *Formally,* which means in a formal manner, should not be confused with *formerly*, which means "in times past" or "previously":

Work on the project was *formally* begun when public funds were made available.
Civil engineers *formerly* did the planning now done by computers.

Former, Latter. *Former* refers to the first named of two, while *latter* refers to the last named of two. *First* and *last* are used to refer to three or more items.

Hanged, Hung. *Hanged* is used for people and *hung* for objects or things:

The negatives were *hung* up to dry.

Half a, A half, A half a. *Half a* and *a half* are in acceptable use, but not *a half a*:

Soldering will require *half a* day.
Soldering will require *a half* day.

Healthful, Healthy. *Healthful* means "giving health." *Healthy* means "having health."

If, Whether. To introduce adverbial clauses expressing a condition, *if* is used. To introduce noun clauses, either *if* or *whether* may be used, though *whether* is preferred:

If it is in the best interest of safety, the operator is free to make the decision.
Whether the shipment arrives or not, the work can proceed on schedule.

In back of. If the meaning is *behind*, it is incorrect to use *in back of*:

The safety switch is located *behind* (not *in back of*) the panel.

Illusion. See *Allusion*.

Imply, Infer. *Imply* means "to hint" or "to suggest," while *infer* means "to draw a conclusion":

The abstract *implied* that controls in the study were inadequate.
On the basis of experience, we *infer* that substance A is greatly superior to substance B.

In, Into. *In* denotes location, *into* direction. *In* should not be used for *into*:

The disc is placed *in* the medium.
The tube may be moved *into* the flame.

In my estimation. This phrase, along with *in my judgment, in my opinion,* is often overworked. It is frequently better to say simply, *I think, I feel, I believe.*

In regards to. Incorrect form of *in regard to.*

Individual, Party. When referring to a single person, use *individual. Party*, except in legal writing, refers to a group:

Many *individuals* (not *parties*) must cooperate to bring the project to completion.
A *party* of electronic technicians visited the atomic installation.

Infer. See *Imply.*

Inside of. When *inside* is used as a preposition, the *of* is unnecessary. When used in reference to time, *inside of* should not be made to take the place of *within*:

The glass was packed *inside* the casing.
Preliminary preparations should begin *within* an hour before the experiment.

Irregardless. No such acceptable word exists for *regardless.*

Is when, Is where. These constructions should be avoided, especially when making definitions:

Maximum stress *is made when* the top unit is brought into locking position.

Its, It's. Forms frequently confused, even by otherwise literate writers. *Its* is the possessive pronoun and never requires an apostrophe before or after the *s. It's* is a contraction of *it is* and always requires an apostrophe.

Kind, Sort. These are singular forms which may be modified by *that* or *this*. *Those* and *these* are used only with plural forms:

This kind of substance is more malleable than the other kinds.
Those kinds of textiles are thought to be flammable.

Later, Latter. *Later* refers to time and is the comparative form of *late*. *Latter* refers to the last named of two:

The first shipment of experimental animals was *late* in arriving, but the second was even *later*.
Of the two methods, the *latter* proved more effective.

Lay, Lie. Today I *lay* it down; yesterday I *laid* it down; many times I have *laid* it down. Today I *lie* down; yesterday I *lay* down; many times I have *lain* down:

The materials still *lay* in the position we had left them yesterday.
They were in the process of *laying* the foundation.
Flammable materials were left *lying* about.

Lead, Led. Today I *lead*; yesterday I *led*; many times I have *led*:

Lead the wire into the aperture before tightening the vise.
The wire was *led* into the aperture.

Learn, Teach. To *learn* is to acquire knowledge. To *teach* is to impart knowledge:

All technicians must *learn* to write effective reports.
Each class must *teach* what it knows to the group that follows.

Leave, Let. *Leave* means "to depart from," while *let* means "to permit." The two words should not be confused:

Leave the area and *let* the dust settle.

Less. See *Fewer*.

Like, As, As if. In formal and technical English, *like* should be used as a preposition and *as* or *as if* as a conjunction:

The specimen looked *like* a checkerboard under the microscope.
The viscosity was *as* (not *like*) it should be.

Likely. See *Apt*.

Likely, Liable. In formal and technical English *likely* means "probable, to be expected." *Liable* means "susceptible to something unpleasant" or "legally responsible":

The statistical analysis is *likely* to show the cause of the side effects.
The test is *liable* to produce several different harmful reactions.

Lose, Loose. The verbs *lose* and *loose* mean "to cease having" and "to set free," respectively. *Loose* as an adjective means "free, not fastened":

You will *lose* more time if you *loosen* the bracket again.
The bracket is *loose* and will not hold.

Lots, Lots of. In formal and technical English, *much* or *a great deal* should be used:

There was *a great deal* of salvage material.
There was *much* electrostatic interference.

Majority. *Majority* is "the quality or state of being greater." The term should not be used with measures of time and distance:

We spent *most* (not the *majority*) of the time waiting for supplies.
In a *majority* of the cases the patients will survive the initial shock.

May. See *Can.*

May be, Maybe. *May be* is a verb form while *maybe* is an adverb meaning "perhaps":

The first assay *may be* sufficient.
Maybe the first assay will be sufficient.

Most. This term should not be used in formal and technical English in place of *almost*:

Almost everybody working on the project agreed about the problem.

Mutual. See *Common.*

Nowhere near, Nowheres near. In formal and technical English it is best to use *not nearly* in place of either of these phrases:

There were *not nearly* (not *nowhere near*) enough albino rats available at the start of the study.

Number. See *Amount.*

On to, Onto. An adverb and a preposition, *on to* refers to forward motion of mind or body. The preposition *onto* refers to getting aboard something:

The lecturer moved *on to* explain the next theorem.
The cylinders were lifted *onto* the conveyor belt with a hydraulic crane.

One another. See *Each other.*

Outside of. In formal and technical English this phrase should not be used for *except* or *besides*:

Nothing remained in the receptacle *except* crystals.

Party. See *Individual.*

Per cent. In technical writing the % sign is used after figures. Both the word and the sign mean "by the hundred." Strictly speaking, *percentage* means "allowance, rate of interest, discount on a hundred":

Loss was reduced by *34%*.
Investment in the new machinery did not offer an attractive *percentage* of return.
Percent may be written as a single word and need not be followed by a period.

Plenty. In formal and technical English, *plenty* should not be used for *very*. *Plenty* is a noun meaning "a great deal" or an "abundance":

It will be *very* (not *plenty*) late before the initial work is completed.
Plenty of nitrate is available for shipment.

Practical, Practicable. *Practical* refers to practice, action, and the useful as opposed to the theoretical. *Practicable* means "capable of being put into practice," that which is feasible:

The most *practical* method of extraction was used.
The most logical plan was the least *practicable*.

Principal, Principle. In all forms of writing these words are confused. The adjective *principal* means "chief, main." The noun *principal* means "leader, chief officer." In finance *principal* means "a capital sum, as distinguished from interest or profit." The noun *principle* means "fundamental truth" or "basic law or doctrine":

Car exhaust is a *principal* cause of air pollution.
Our high school *principal* was interested in technical education.
The new process called for a sizeable investment of *principal* in the form of equipment.
Chlorination is based on an established chemical *principle*.

Proposition, Proposal. *Proposal* means "an offer, a scheme, a plan, or a bid." *Proposition* has more to do with an idea, statement, or principle set forth for discussion.

What is your *proposal* to improve communication between the two groups?
They advanced two new *propositions* for study.

Raise, Rise. Today I *raise* the flag; yesterday I *raised* the flag; many times I have *raised* the flag. Today I *rise* from bed; yesterday I *rose* from bed; many times I have *risen* from bed:

The enamel *rose* from the heat.
They *raised* the level of production.

Real. In formal and technical English *real* should not be used as an adverb in place of *very* or *extremely*:

The experiment proved *very* (not *real*) useful.
The laboratory assistant was *extremely* (not *real*) pleased with the result.

Reason is because. This construction is awkward and illogical:

The *reason* for the delay *is* bad weather.

Respectfully, Respectively. *Respectfully*, means "in a manner showing respect"; *respectively* means "each in the order given":

The Committee treated *respectfully* the requests of both working groups.
The Committee dealt with the requests of each petitioning group *respectively*, as shown in the accompanying minutes.

Right along. This expression is not appropriate in formal and technical writing:

Plant construction moved forward without interruption (not *right along*.)

Rough. In formal and technical English *rough* is not acceptable as a substitute for *difficult* or *unpleasant*:

It was a *difficult* (not *rough*) experiment to perform.

Run, Ran. In technical English *run* is acceptable in the sense of "perform" or "execute" when one is speaking of experiments or work in progress. But in other uses *run* should not be used in place of "operate," "manage," or "supervise":

We *ran* the experiment a third time, to verify our earlier data.
The quality control section is *managed* (not *run*) well.

Shall, Will, Should, Would. To express conditions and obligations, *should* is used in all persons. To express a wish or customary action, *would* is used in all persons:

If the pressure *should* rise too rapidly, check the safety valve.
The technician *should* follow the manual.
He *would* always proceed in the same fashion.

Shape. In technical and formal English, do not use *shape* for *condition*:

Spinning machines are in good *condition* (not *shape*) for immediate use.

Show up. In formal and technical English *show up* is not acceptable for "arrive" or for "discover, reveal, expose, or appear":

The shipment did not *arrive* (not *show up*) on schedule.
The error did not *appear* (not *show up*) the first time we studied the data.

Sight. See *Cite*.

Simulate, Dissimulate. To *simulate* is "to assume the appearance of," to practice an operation, for example, as if it were real. To *dissimulate* implies pretending for questionable or dishonest purposes:

The pilots *simulate* emergency landing procedures before taking off.
The explanation of the breakdown seemed to *dissimulate* the true cause.

Site. See *Cite.*

So. Much overworked and misused. *So* should not be used to join main clauses. In clauses of purpose, *so that* is needed:

Pressure was inadequate, *and so* we increased the volume. or, *Because* pressure was inadequate, we increased the volume. or, We increased the volume *so that* pressure would be adequate.

Sort. See *Kind.*

Sure. *Sure* should not be used in formal and technical English as a substitute for *surely* or *certainly*:

The new extraction technique *surely* (not *sure*) was a success.

Their, There, They're. It is a serious error to confuse these forms in technical writing. *Their* is a possessive pronoun; *there* is an adverb or expletive; *they're* is a contraction of *they are* but is not used in formal and technical writing:

Their rejection of the plan caused difficulty.
There will be discussion again tomorrow.

There, They're. See *Their.*

Through. Use *finished* in formal and technical English:

The preliminary work is *finished* (not *through*).

To, Too, Two. Confusion of these words is common, but it is a serious mistake. *To* is a preposition; *too* is an adverb; and *two* is a numeral:

Follow the diagram *to* the right.
Cooling occurred *too* rapidly.

Try and. The formal phrase is *try to*:

The foreman must next *try to* remove the secondary outlet.

Unique. In formal and technical English one cannot logically compare *unique* since it means "single in kind or excellence":

This method of flocculation is *unique* (not *more unique*).

Way, Ways. In formal and technical English *way* should not be used to mean "away." *Ways* should not be confused with *way*:

The other connection may be seen *away* across the diagram. A short *way* from the chute is the incinerator.

Whether. See *If*.

Which, that; Who, that. *Which* refers to things; *who* refers to people. *That* refers to either things or people, often in restrictive clauses:

The heat ducts, *which* constantly expanded, caused delays for adjustment.
The heat ducts *that* expanded caused delays for adjustment.
The chief accountant is the one *who* complained.
The chief accountant is the one *that* complained.

Who, That. See *Which, That*.

-Wise. In good formal and technical English, this suffix should simply be avoided. Its common use colloquially has become a joke:

In terms of weight and light (not *weight-wise* and *light-wise*) the distillate was perfect, but the color (not *color-wise*) was not satisfactory.

THE TECHNICIANS' GUIDE TO GRAMMATICAL TERMS

By the time a person undertakes to do technical writing his habits of grammar will have become fixed and largely a matter of unconscious routine. One will write according to his practice and without full attention to grammatical considerations.

A course in technical writing may well be the only writing course taken in college by those studying for the sciences and technologies. Many such students may find they are rusty on grammar and grammatical terminology, a subject taken up earlier in one's education. Nonetheless, instructors of technical writing—as well as instructors of technical courses—will frequently call grammatical errors to the attention of student writers.

This Guide offers a compact, greatly abbreviated listing—with definitions and examples—of many of the most commonly used grammatical terms. Only the most unusual basic technical writer will not be obliged to refer at some time to the Guide for help.

Absolute. Any *absolute* element is grammatically independent of the remainder of the sentence in which it appears but is connected by thought. The absolute phrase generally consists of a noun or pronoun in the subjective case followed by a participle. Such a construction may also be called *nominative absolute*:

The first experiment completed, we waited for further instruction.
The second tier of girders was installed, *the supporting elements left protruding around all parameters.*

Active Voice. See *Voice*.

Adjective. A word that describes or limits the meaning of a noun, pronoun, or gerund. *Descriptive adjectives* isolate a particular quality of an object: *glass* beaker, *purple* disc, *turbid* water. *Limiting adjectives* restrict the meaning of a noun to a particular object or indicate quantity or number. Five kinds of limiting adjectives are:

numerical: two furnaces, *third* layer; *articles: the* report, *an* error, *a* technician; *interrogative: which* office?; *what* question?; *demonstrative: that* incinerator, *this* section; *possessive: our* table, *their* plant.

Adjective Clause. A subordinate, or dependent, clause used as an adjective:

The laboratory assistants *who work here* are well trained.
Men *who complained of delays* were given an explanation.

Adverb. A word that describes or limits the meaning of a verb, an adjective, or another adverb. According to meaning, adverbs show:

Place: Store the plasma *inside*. (modifies verb)
Time: The chambers were *always* cold. (modifies adjective)
Manner: The mice were *aggressively* hostile. (modifies adjective)
Degree: Work was *very* quickly started. (modifies adverb)

Adverb Clause. A subordinate, or dependent, clause used as an adverb:

When you complete the first phase, cover the top layer with foam. (modifies verb *cover*)

The coating peeled *where the heat was greatest*. (modifies verb *peeled*)

Adverbial Noun. A noun used as an adverb (also called *adverbial objective*):

The overhang extended a *foot*.
The pigeon flew *home* rapidly.

Agreement. Correspondence in person and number between a subject and verb; in person, number, and gender between a pronoun and its antecedent; and in number between a demonstrative adjective and its noun:

Agreement of verb and subject:
I *have*; you *have*; he, she, it *has*; we *have*; you *have*; they *have*.

Agreement of pronoun and antecedent:
Water seeks *its* level.
The *nurses* scrubbed *their* hands.
Each *man* must make *his* judgment.

Agreement of *demonstrative adjective* and its noun:
this room; these bottles; that sound

Antecedent. The word or group of words to which a pronoun refers:

A *technician* who writes well is always in demand. (antecedent of *who*)
Always be prepared is still as useful a caution as it used to be. (antecedent of *it*)

Appositive. A substantive placed beside another substantive and denoting the same person or thing:

The new plastic, *a malleable substance*, has many uses. (in apposition with *plastic*)
The objective—*to develop an odorless fuel*—is clear. (in apposition with *objective*)
The argument *that he presented* was convincing. (in apposition with *argument*)

The first two appositives above are *nonrestrictive* and are set off with commas or dashes. The last appositive is *restrictive* and is not set off with commas.

Article. The definite article is *the*. The indefinite articles are *a* and *an*. All are used as adjectives.

Auxiliary. A verb form used to help (a *helping* verb) form the voices, moods, and tenses of other verbs. The common auxiliaries are *be* (in its various forms), *have, shall, will, should, would, may, can, might, could, must, ought,* and *do*:

The project *ought* to be completed on time, but no one is able to say if it *will* be.

Case. The inflectional forms of nouns and pronouns that indicate their relationships to other words in the sentence. The three cases in English are *subjective* (nominative), *possessive* (genitive), and *objective* (accusative). The possessive is the only distinctive case form of the noun remaining in English. The subjective and objective cases of nouns are determined by relation to the verb or preposition:

The Noun

	Singular		*Plural*	
Subjective	chemist	cherry	chemists	cherries
Possessive	chemist's	cherry's	chemists'	cherries'
Objective	chemist	cherry	chemists	cherries

The Pronoun
Singular

	First Person	*Second Person*	*Third Person*
Subjective	I	you	he, she, it
Possessive	my, mine	your, yours	his, her, hers, its
Objective	me	you	him, her, it

The Pronoun
Plural

	First Person	*Second Person*	*Third Person*
Subjective	we	you	they
Possessive	our, ours	your, yours	their, theirs
Objective	us	you	them

Clause. A group of words containing a subject and verb. There are two kinds of clauses:

Main Clause: (also called *independent* or *principal* clause) Makes an independent assertion and stands alone—it does not function as part of another clause and does not modify a word outside of itself:

When the final truck entered, *the watchman locked the gate.*

Subordinate Clause: (also called *dependent* clause) Does not make an independent assertion that stands alone—functions as part of another clause or modifies a word outside itself. Subordinate clauses as grammatical units function as adjectives, adverbs, or nouns:

When the final truck entered, the watchman locked the gate. (modifies verb of main clause—functions as adverb)
The technician *who works carefully* will gain approval. (modifies subject of main clause—serves as adjective)
What you are asking is not clear. (serves as subject of sentence—functions as noun)

Collective Noun. See *Noun.*

Comma. An internal mark of punctuation used properly to separate or set off elements within a sentence. Many technical writers use commas illogically. To avoid this, follow these four general rules:

1. The comma should be used to separate two independent clauses joined by a coordinating conjunction (and, but, or, nor, for, yet):

The acid bath was prepared, but the temperature required adjustment.

2. The comma should be used to separate a long introductory clause or phrase from the rest of the sentence:

Because of the accelerated production schedule, some technicians were asked to work overtime.

3. The comma should be used to set off nonrestrictive, parenthetical, interrupting elements from the rest of the sentence:

The main structural support, which had been designed to bear additional stress, was hoisted into position.

4. The comma should be used to set off two or more coordinating words or sentence elements modifying the same noun:

The green, colloidal matter tended to pollute the brackish, grey liquid material at the bottom of the receptacle.

Common Noun. See *Noun*.

Comma Splice and Fused Sentence. A comma splice is the linking of two independent clauses with only a comma between them. A fused sentence, even worse, runs two independent clauses together without any punctuation or coordinating conjunction between them. There are three legitimate ways to avoid a comma splice or fused sentences in the joining of two independent clauses: 1) By a coordinating conjunction; 2) by a coordinating conjunction and comma; or 3) by a semicolon with or without a coordinating conjunction.

Comma Splice: All the samples are collected, the filtrating device is working well.
Fused Sentences: All the samples are collected the filtrating device is working well.
Corrected: All the samples are collected, and the filtrating device is working well.
Corrected: All the samples are collected; the filtrating device is working well.

Depending on the relationships between the ideas, it is also possible to avoid a comma splice or fused sentences by subordinating one sentence element to the other:

Corrected: Because the filtrating device was working well, all the samples are collected.

Comparison of Adjectives and Adverbs. Denotes changes in the forms of adjectives and adverbs to indicate degrees of quality, quantity, or relation. There are three degrees of comparison: *positive*—the simple or basic form of the adjective or adverb; *comparative*—used in comparing two units to express a higher degree than that denoted by the simple form; *superlative*—used in comparing more than two units to indicate the highest amount of that denoted by the simple form.

Simple	*Comparative*	*Superlative*
hot	hotter	hottest
close	closer	closest
effective	more effective	most effective
inefficient	more inefficient	most inefficient

Complement. A term used to describe the word or group of words that completes the meaning of a verb. Complements may be of six different kinds: *direct object, indirect object, predicate noun* (or *predicate nominative*), *predicate adjective, objective complement,* or *retained object.*

The scientists built a *reactor*. (direct object)
The company gave *employees* a bonus. (indirect object)
Harry is an excellent *chemist*. (predicate noun)
The surface felt *rough*. (predicate adjective)
The biologists named Jack their *chairman*. (objective complement)
Our research team was awarded a *prize*. (retained object)

Conjugation. A term used to describe the changes in the inflectional forms of a verb to show tense, voice, mood, person, and number. Given below is a complete conjugation table of the verb *to see* in all its tenses, voices, moods, persons, and numbers. The table may serve as a guide to the presentation of other verbs in their correct forms.

CONJUGATION OF THE VERB *to see*

Principal Parts: see, saw, seen

Indicative Mood

Active Voice		Passive Voice	

Present Tense

Singular	*Plural*	*Singular*	*Plural*
I see	we see	I am seen	we are seen
you see	you see	you are seen	you are seen
he (she, it) sees	they see	he is seen	they are seen

Past Tense

I saw	we saw	I was seen	we are seen
you saw	you saw	you were seen	you are seen
he saw	they saw	he was seen	they are seen

Future Tense

I shall see	we shall see	I shall be seen	we shall be seen
you will see	you will see	you will be seen	you will be seen
he will see	they will see	he will be seen	they will be seen

Present Perfect Tense

I have seen	we have seen	I have been seen	we have been seen
you have seen	you have seen	you have been seen	you have been seen
he has seen	they have seen	he has been seen	they have been seen

Past Perfect Tense

I had seen	we had seen	I had been seen	we had been seen
you had seen	you had seen	you had been seen	you had been seen
he had seen	they had seen	he had been seen	they had been seen

Future Perfect Tense

I shall have seen	we shall have seen	I shall have been seen	we shall have been seen
you will have seen	you will have seen	you will have been seen	you will have been seen
he will have seen	they will have seen	he will have been seen	they will have been seen

Conditional Mood

Present Tense

I may see	we may see
you may see	you may see
he may see	they may see

Present Perfect Tense

I may have seen	we may have seen
you may have seen	you may have seen
he may have seen	they may have seen

Past Tense

I might (should, would) see	we might (should, would) see
you might (should, would) see	you might (should, would) see
he might (should, would) see	they might (should, would) see

Subjunctive Mood

	Active Voice	Passive Voice
	Present Tense	
Singular:	if I, you, he see	if I, you, he be seen
Plural:	if we, you, they see	if we, you, they be seen
	Past Tense	
Singular:	if I, you, he saw	if I, you, he were seen
Plural:	if we, you, they saw	if we, you, they were seen
	Present Perfect Tense	
Singular:	if I, you, he have seen	if I, you, he have been seen
Plural:	if we, you, they have seen	if we, you, they have been seen
	Past Perfect Tense	
Singular:	if I, you, he had seen	if I, you, he had been seen
Plural:	if we, you, they had seen	if we, you, they had been seen

Imperative Mood

Present Tense

see	be seen

Infinitives

Present Tense

to see	to be seen

Perfect Tense

to have seen	to have been seen

Participles

Present Tense

seeing	being seen

Past Tense

seen	been seen

Perfect Tense

having seen	having been seen

Gerunds

Present Tense

seeing	being seen

Perfect Tense

having seen	having been seen

Progressive Forms, Indicative Mood

Present Tense

I am seeing, etc.	we are being seen, etc.

Past Tense

I was seeing, etc.	we were seeing, etc.

Progressive Forms, Subjunctive Mood

Present Tense

if I be seeing, etc.	if we be seeing, etc.

Past Tense

if I were seeing, etc.	if we were seeing, etc.

Conjunction. A word used to connect other words or groups of words, phrases, and clauses. There are two kinds of conjunctions, *coordinating* and *subordinating*. The *coordinating* conjunctions are: *and, but, or, nor, for* and, sometimes, *yet*; these are used to join words, phrases, and clauses of coordinate or equal grammatical rank. The subordinating conjunctions are: *after, as, because, if, when, although,* and others, as the construction may demand; these are used to join subordinate grammatical units to main clauses:

We failed to complete the project on time, *for* we had not adhered carefully to our work schedule.
Although there was continued willingness to cooperate, a more equitable division of labor was needed.

Conjunctive Adverb. An adverb used as a connective. These words, however, are properly used as adverbs and not as conjunctions: *also, further, however, nevertheless, hence,* and *consequently*. They are best thought of as nonrestrictive modifiers or interrupting elements and in no case should they be employed as conjunctions—to do so is to create the serious grammatical error of fusing or splicing independent sentence elements in the absence of legitimate conjunctions. Depending on the position of the conjunctive adverb, commas or endmarks of punctuation (semi-colon or period) must be used with it:

There was no progress in the talks; *nevertheless*, both sides agreed to continue negotiating.
Consequently, all the furnaces were kept going.
There was, *also*, the issue of administrative control.
They had promised to make the delivery of steel on Monday; *however*, the snow storm made that impossible.

Correlative Conjunctions. Conjunctions used in pairs to express a mutual, equal relationship between words, phrases, and clauses. They are *either...or; neither...nor; not only... but also*; and *both...and*.

Declension. See *Case*. Declension refers to the change of form in nouns and pronouns to show their number and case.

Ellipsis. The omission of a word or words necessary to complete a sentence grammatically, the meaning of which the reader assumes readily from the context:

We completed our survey before they (*completed theirs*). (Underscored words, readily understood by reader or listener, may be elliptically omitted.)
As a researcher, he is more experienced than I. (*am*).

Expletive. A word such as *it* or *there* used to introduce a sentence in which the subject follows the verb. The expletive at the start of the sentence is a grammatical stand-in for the true subject:

There was *sediment* in the tank. (*Sediment* is the true subject; *there* is an expletive.)
It is a logistical *problem* to be solved. (*Problem* is the true subject; *it* is an expletive.)

Gerund. See *Verbal*.

Independent Element. See *Absolute* and *Interjection*.

Infinitive. See *Verbal*.

Inflection. See *Case* and *Conjugation*.

Interjection. An exclamatory word or words used to express emotion. It is grammatically independent of the rest of the sentence. (The interjection is seldom if ever used in technical writing.)

Wow! What a spectacular sight!
Ouch! I burned my finger with that lighter!

Linking Verb. See *Verb*.

Mood. (See *Conjugation.*) The form of the verb indicating the speaker's attitude toward what he says. There are three moods in English: *indicative, imperative,* and *subjunctive.*
The *indicative* mood states a fact or asks a question:

The shipment is here. Is the package ready?

The *imperative* mood expresses a command or request:

Report this to the office. Give me your attention.

The *subjunctive* mood expresses a condition contrary to fact, a wish, a doubt, or a resolution:

The experimental mice look as if they are dying. I wish it were possible to save them.

Nonrestrictive Modifier. See *Restrictive Modifier.*

Noun. A word that names a person, place, or thing.
A *common* noun names any one of a class of persons, places, or things:

technician, wheel, bracket

A *proper* noun names a specific person, place, or thing.

John F. Kennedy, Nova Scotia, Apollo 7

A *concrete* noun names something that can be perceived by the senses.

coal, steel, echo, sulfur dioxide

An *abstract* noun names something not perceived by the senses.

science, research, maturity

A *collective* noun names a group by using a singular form.

committee, team, audience

Number. (See *Conjugation, Noun, Pronoun,* and *Verb*.) The form of a noun, pronoun, verb, or demonstrative adjective that indicates one (singular) or several (plural) persons or things.

Object. A substantive that directly or indirectly receives the action of a transitive verb. The object of a preposition is the substantive that follows the preposition:

Direct object of verb: He dropped the *beaker.*
Indirect object of verb: The committee sent *us* word of the decision.
Object of preposition: The pathologist placed the tissue under the *microscope.*

Objective Complement. See *Complement.*

Parenthetical Expression. (See *Comma, Interjection* and *Restrictive and Nonrestrictive Modifiers*.) An inserted expression that interrupts the thought of a sentence. Parenthetical elements are set off by commas, dashes, or parentheses:

The failure of the experiment, it was generally believed, brought the project to a dead end.
His excuse—and this was no surprise—impressed nobody.
The chairman (formerly the secretary) rose to speak.

Participle. See *Verbal.*

Parts of Speech. The classification of words according to their function in a sentence. The eight parts of speech in English are: *adjective, adverb, conjunction, interjection, noun, preposition, pronoun,* and *verb*. (Each of these is listed separately in this *Guide*.)

Passive Voice. See *Voice.*

Person. (See *Conjugation* and *Pronoun*.) The form of a pronoun and verb used to indicate the speaker: *first person*—I am; the person spoken to: *second person*—you are; or the person spoken about: *third person*—he is.

Phrase. (See *Clause*.) A group of words functioning as a grammatical unit but not containing a subject or a verb. On the basis of their form and function, phrases are classified as:

Prepositional: The biologist placed the specimen *in the dish*. (used as adverb to modify *placed*)
The chief spokesman *for the new idea* was Mr. Kowalski. (used as adjective to modify *spokesman*.)
Participial: The man *running the experiment* is a nuclear scientist. (adjective)
Gerund: *Scrubbing the equipment* is always a chore. (noun)
Infinitive: *To see a team working as a unit* is always reassuring. (noun)
Verb: He *has been consulting* with the central office.

Predicate. (See *Verb*.) The part of the sentence that makes a statement about the subject. The predicate consists of the verb and its complement and any modifiers:

Clear technical writing *is absolutely necessary to communicate important information.*

Preposition. A word used to relate a noun or pronoun to some other word in the sentence:

Test animals were placed *in* the cages.
Remove the tiles *from* the oven.
Activity occurred *outside* the window.

Principal Clause. (See *Clause*.) A main or independent clause.

Principal Parts of a Verb. (See *Conjugation*.) The three forms of a verb with which one may construct all the other tenses and forms of the verb. They are the *present infinitive*, the *past tense*, and the *past participle*:

write	wrote	written
study	studied	studied
grow	grew	grown

Pronoun. A word that takes the place of a noun. The noun for which a pronoun stands is called the *antecedent* (see listing). The classes of pronouns are: *indefinite, intensive, interrogative, personal, reciprocal, reflexive,* and *relative.*

Indefinite Pronouns: These do not refer to a definite or particular person, place, or thing: *all, any, anybody, each, few, one, everyone, something.*
Intensive Pronouns: These emphasize the substantive to which they refer: *myself, yourself, himself, herself, itself, oneself, yourselves,* and *themselves.*
Interrogative Pronouns: These are used to ask questions: *who, which, what, whoever,* and *whatever.*
Personal Pronouns: These pronouns make clear by their forms either the person speaking (first person), the person spoken to (second person), or the person or thing spoken of (third person): *I, we; you; he, she, it; they.* See *Case* listing.)
Reciprocal Pronouns: These are used to show interaction and occur only as the object of a verb or preposition: *each other, one another.*
Reflexive Pronouns: These refer to the subject of a verb or verbal and have the same forms as the *intensive* pronouns above:

They will go by *themselves.*
The chemist bruised *himself* while setting the dial.

Relative Pronouns: These are used to connect a subordinate clause to an independent clause: *who, which, that.*

Restrictive and Nonrestrictive Modifiers. Restrictive modifiers—a word, phrase, or clause—restrict or identify the word they modify. According to the meaning of the writer and the syntax of the sentence, a restrictive modifier restricts the meaning of the modified element and, as such, cannot be removed from the sentence without changing the essential sense. A nonrestrictive modifier adds additional helpful information, but it does not absolutely restrict or identify the element it modifies; a nonrestrictive modifier could be removed without altering the essential meaning of the sentence. Restrictive modifiers are not set off with commas; nonrestrictive modifiers are set off with commas.

Restrictive Modifiers: Technicians *who are trained in mathematics* are needed in the national space program.
The epidermis tissue *on the right forearm* presented a different texture from that on the left forearm.
Hydrogen sulphide *as it occurs in mineral water* may be flammable and poisonous.
Nonrestrictive Modifiers: Hydrogen sulphide, *found in many mineral waters*, has a disagreeable odor.
Space technology, *unheard of 25 years ago*, is a rapidly expanding field.
The liquid in the tube, *of a brownish rusty hue*, was unaffected by the carbon filter.

Sentence. A group of words expressing a thought and containing a subject and a verb.

A *simple* sentence has one main clause. The subject or the verb may be compound:

Technology advances.
Students and professors worked together.

A *compound* sentence has two or more main clauses:

The project continued, but progress was slow.

A *complex* sentence has one main clause and one or more subordinate clauses:

The men in the control center were elated when they learned that the moon landing was successful.

A *compound-complex* sentence has two or more main clauses and one or more subordinate clauses:

By January the diverse teams completed their preliminary investigations, but, while they kept their equipment at the ready, there was no real work accomplished beyond that date.

According to their main intention and use, sentences are further classified as declarative, interrogative, imperative, and exclamatory.

A *declarative* sentence states or asserts something:

The work began. Construction continued.

An *interrogative* sentence asks a question:

How much will you earn? Is it interesting work?

An *imperative* sentence requests or commands:
Handle this package with care. Please work quickly.

An *exclamatory* sentence expresses great emotion and is followed by an exclamation point.

What confusion! They won the contest!

Sentence Fragment. A group of words set up and punctuated as if it were a sentence but which is grammatically incomplete. Fragments take many forms but generally appear as subordinate clauses, phrases, appositives, or a substantive followed by a phrase or subordinate clause. None of these elements can stand alone as a sentence; to attempt to make them do so may be taken as a sign of illiteracy.

Subordinate clause as a fragment: The new conveyor device reduced time-input substantially. *Although production was still down.*
Corrected: The new conveyor device reduced time-input substantially, although production was still down.
Phrase as fragment: Spray the exterior lightly with lacquer. *After three hours.*
Corrected: Spray the exterior lightly with lacquer after three hours.

Appositive as fragment: It was an improved, automated system. *A system for rapid retrieval of information.*
Substantive followed by a phrase or clause as fragment: The ecologist gaining the greatest support for his views.
Corrected: Smith is the ecologist who gained the greatest support for his views.

Subject. The person or thing about which the predicate of a sentence or clause makes an assertion. The subject word or words is called a substantive. As there may be a compound verb, so may there be a compound subject:

The *statistics* from the government and the *reports* from industry suggested an increase in business activity.

Subjunctive. (See *Conjugation* and *Mood*.) The mood of the verb which expresses a condition contrary to fact, a wish, a doubt, or a resolution.

Substantive. A word or a group of words used as the equivalent of a noun. *Nouns, pronouns, infinitives, gerunds,* and *noun clauses* may serve as substantives.

Syntax. The putting together of words into phrases, clauses, and sentences. The order or arrangement of words to express a thought and expose mutual relationships in a sentence.

Tense. (See *Conjugation*.) The time or the state of action expressed by a verb. Tense is the verb property that relates an action or state of being to time.

Verbal. A grammatical form derived from a verb but which cannot stand for or be used in place of a verb. The three kinds of verbals are the *gerund* (verbal noun), the *participle*, and the *infinitive*. The gerund serves as a noun (substantive), the participle as an adjective, and the infinitive as a noun, adjective or adverb.
Gerunds end in *-ing* and have the function of a noun, such as subject or object of a verb:

Dyeing is a process used in several technologies.
The oceanographers enjoyed *diving.*

Participles serve as adjectives but, being derived from verbs, may take an object. As an adjective, the participle may modify a noun or pronoun. Present participles end in *-ing;* past participles end in *-d, -ed, -t, -n, -en,* or change the form of the vowel:

The chemist *wearing the white jacket* is the director.
The rocks *found in the bottom* stratum were the most ancient.

Infinitives are usually preceded by *to* and are used as nouns, adjectives, or adverbs:

To experiment is a necessity.
There was much work *to do.*
The director went *to inspect* the project.

Verb. A word or group of words which asserts action or indicates state of being. One may speak of transitive and intransitive verbs, regular and irregular verbs, and of linking verbs. (See *Conjugation.*)

Transitive verbs: These require an object to complete their meaning. The verb transfers an action to its object from the subject:

The *astronaut placed* a *flag* on the moon.

Intransitive verbs: These do not require an object to complete their meaning.

Organisms *swam* under the microscope.

Regular verbs: A verb of this kind (also called *weak*) forms its past tense and past participle by adding *-ed, -d,* or *-t* to its present tense: work, work*ed*, work*ed*; experiment, experi*-*ment*ed*, experiment*ed*; deal, deal*t*, deal*t*.

Irregular verbs: A verb of this kind (also called *strong*) does not form its past tense and participle as in the regular verb above. Irregular verbs usually form their past tense and past participles by undergoing an internal change. (Under *Conjugation* see all of the inflected forms of the irregular verb *to see*, the principal parts of which are: *see, saw, seen.*

Linking verbs: These verbs link or join the subject to a noun or adjective that explains, defines, or describes the subject. The verb *to be* in all of its forms is a linking verb (also called *copulative* verb). Other linking verbs include those having to do with sensory perceptions: *look, smell, feel, hear, taste, see, seem, appear, become.*

Voice. The property of a verb that indicates whether the subject does the acting or is acted upon. The *active* voice indicates that the subject is the doer of the action. The *passive* voice indicates that the subject is acted upon.

Active voice: The technician carried out the testing procedure.
Passive voice: The testing procedure was carried out by the technician.
Active voice: The personnel office sent out new regulations for hiring.
Passive voice: New regulations for hiring were sent out by the personnel office.